James J. Buckley
Esfandiar Eslami

An Introduction to Fuzzy Logic and Fuzzy Sets

With 50 Figures
and 14 Tables

Physica-Verlag
A Springer-Verlag Company

Professor James J. Buckley
University of Alabama at Birmingham
Mathematics Department
Birmingham, AL 35294
USA
buckley@math.uab.edu

Professor Esfandiar Eslami [1]
Shahid Bahonar University
Department of Mathematics
Kerman
Iran
eslami@arg3.uk.ac.ir
eslami@math.uab.edu

[1] Thanks to the University of Shahid Bahonar, Kerman, Iran for financial support during my sabbatical leave at UAB. Thanks to the University of Alabama at Birmingham for producing a good atmosphere to do research and teaching. Special thanks to Prof. James J. Buckley for his very kind cooperation that made all possible.

ISSN 1615-3871
ISBN 3-7908-1447-4 Physica-Verlag Heidelberg New York

Cataloging-in-Publication Data applied for
Die Deutsche Bibliothek – CIP-Einheitsaufnahme
Buckley, James J.: An introduction to fuzzy logic and fuzzy sets: with 14 tables / James J. Buckley; Esfandiar Eslami. – Heidelberg; New York: Physica-Verl., 2002
(Advances in soft computing)
ISBN 3-7908-1447-4

Physica-Verlag Heidelberg New York
a member of BertelsmannSpringer Science+Business Media GmbH

Printed in Germany

Softcover Design: Erich Kirchner, Heidelberg

SPIN 10857116 88/2202-5 4 3 2 1 0 – Printed on acid-free paper

To Julianne and Mehra

Helen, Pooya, Peyman and Payam

Contents

Chapter 1

Introduction

This book is intended to be an undergraduate introduction to the theory of fuzzy sets. We envision, sometime in the future, a curriculum in fuzzy systems theory, which could be in computer/information sciences, mathematics, engineering or economics (business, finance), with this book as the starting point. It is not a book for researchers but a book for beginners where you learn the basics.

This course would be analogous to a pre-calculus course where a student studies algebra, functions and trigonometry in preparation for more advanced courses. Chapters 3 through 11 are on fuzzy algebra, fuzzy functions, fuzzy trigonometry, fuzzy geometry, and solving fuzzy equations. However, after this course the student doesn't go on to calculus but to more specialized courses in fuzzy systems theory like fuzzy clustering, fuzzy pattern recognition, fuzzy database, fuzzy image processing and computer vision, robotics, intelligent agents, soft computing, fuzzy rule based systems (control, expert systems), fuzzy decision making, applications to operations research, fuzzy mathematics, fuzzy systems modeling, etc. Therefore, very little of most of these topics are included in this book.

There are many new topics included in this book. Let us point out some of them here: (1) mixed fuzzy logic (Section 3.5); (2) three methods of solving fuzzy equation/problems (Chapter 5); (3) solving fuzzy inequalities (Chapter 6); (4) inverse fuzzy functions (Section 8.5); (5) fuzzy plane geometry (Chapter 9); (6) fuzzy trigonometry (Chapter 10); and (7) fuzzy optimization based on genetic algorithms (Chapter 16).

As a text book similar to a pre-calculus course the sections are short followed by a list of exercises and no references are given. The problems at the end of each section may be divided into three groups: (1) easy (just put the numbers, or data, into the correct equation or formula); (2) medium (determine the correct equation/formula for the data or give an argument for some result stated in the text); and (3) hard.

These problems are all mixed up and we do not indicate their difficulty.

Many students would not try a problem if they were told it was a hard problem. The "hard" problems are not difficult because they are of an advanced nature; they are "elementary" hard problems since they are easy to state in an introductory course. There are not many of these hard problems but there are a few that could warrant publication if solved.

Prerequisites are algebra, trigonometry and elementary differential calculus. We do take derivatives, and some partial derivatives, but these are all elementary and can be taught within the course. In one place, Section 4.6 , we have an integral, and its evaluation can easily be shown by introducing the fundamental theorem of calculus.

It is difficult, in a book with a lot of mathematics, to achieve a uniform notation without having to introduce many new specialized symbols. So what we have done is to have a uniform notation within each section. What this means is that we may use the letters "a" and "b" to represent a closed interval $[a, b]$ in one section but they could stand for parameters in an equation in another section.

We will have the following uniform notation throughout the book:

(1) we place a "bar" over a letter to denote a fuzzy set ($\overline{A}$, $\overline{F}$, etc.);

(2) an alpha-cut is always denoted by "α";

(3) fuzzy functions are denoted as $\overline{F}$, $\overline{G}$, etc.; and

(4) $\mathbf{R}$ denotes the set of real numbers.

The term "crisp" means not fuzzy. A crisp set is a regular set and a crisp number is a real number. There is a potential problem with the symbol "$\leq$". It usually means "fuzzy subset" as $\overline{A} \leq \overline{B}$ stands for $\overline{A}$ is a fuzzy subset of $\overline{B}$ (defined in Chapter 3). However, in Section 4.5 $\overline{A} \leq \overline{B}$ means that fuzzy set $\overline{A}$ is less than or equal to fuzzy set $\overline{B}$. The symbol "$\leq$" , unless we are in Section 4.5, will mean "fuzzy subset" except when we explicitly state otherwise.

The basic material is in Chapters 3 through 11. Then you can choose from different topics in Chapter 2 or from Chapters 12 through 16. Chapter 3 introduces fuzzy sets , their α-cuts and the algebra of fuzzy sets.

Fuzzy numbers and the arithmetic of fuzzy numbers is in Chapter 4. There are many properties of the real numbers we use frequently that need to be translated over to fuzzy numbers. These include: (1) finding the distance between two real numbers ($|x - y|$), becoming the distance between fuzzy numbers in Section 3.7, (2) finding the max and min of two real numbers, resulting in fuzzy max and fuzzy min in Section 4.4, and (3) having inequalities between real numbers ($x \leq y$), extended to inequalities between fuzzy numbers in Section 4.5.

Something special to fuzzy systems theory, defuzzification, is discussed in Section 4.6. Then we go to solving fuzzy equations and fuzzy inequalities in

Chapters 5 and 6. Fuzzy relations, including transitive closure, fuzzy equivalence relations and solving fuzzy relational equations, comprises Chapter 7.

Functions are of great importance in mathematics and fuzzy functions are important in fuzzy systems theory. How we usually get fuzzy functions, via the extension principle or α-cuts and interval arithmetic, is studied in Chapter 8. But also in Chapter 8 we look at types of fuzzy functions, inverse fuzzy functions and elementary derivatives of fuzzy functions.

The geometry of the plane is central to mathematics so we introduce fuzzy geometry (fuzzy circles, fuzzy lines, etc.) in Chapter 9.

To finish off the "basics" we have fuzzy trigonometry and solving systems of fuzzy linear equations in Chapters 10 and 11, respectively.

Discrete possibilities with applications to fuzzy Markov chains are discussed in Chapter 12. Chapter 14 contains approximate reasoning with blocks of fuzzy rules. So called " soft computing", including (fuzzy) neural nets, fuzzy sets and genetic algorithms is included in Chapters 13 and 15.

Fuzzy optimization, based on genetic algorithms is in the last chapter.

For completeness we have also added the basics of crisp and fuzzy logic in Chapter 2.

We suggest the material to cover in a one semester course is Chapters 3, 4, 5, 7 and 8, and the choose from the remainder to finish out the term.

There are a couple of ways to expand the book to surely fill up a two semester course. The students can download a training algorithm for layered, feedforward, neural nets in Chapter 13 and then do the training needed to fully complete the exercises in Section 13.2.1 of that chapter. Also, the students can download genetic algorithm software in Chapter 15 to use to complete the fuzzy optimization problems (get numerical answers) given in the exercises in Chapters 15 and 16.

Being very optimistic, we hope to have a second edition of this book. So, please send us your comments on what material needs to be covered and especially you favorite problems. If we include your problem in a second edition we will reference you, by name, to be attached to the problem.

Chapter 2

Logic

2.1 Introduction

This chapter is a brief introduction to fuzzy logic. To see how fuzzy logic extends and generalizes classical logic we start in the next section with propositional logic. Propositional logic deals with finding the truth values of formulas containing atomic propositions, whose truth value is either zero or one, connected by "and" ($\wedge$), "or" ($\vee$), implication ($\rightarrow$), etc.

Then, in the third section we review the basic results from crisp set theory. Also in the third section we point out the identification between set theory and propositional logic.

The beginning of fuzzy logic follows in the fourth section where:

(1) we allow truth values to be any number in $[0, 1]$, and

(2) we find truth values of formulas using a t-norm for $\wedge$ and a t-conorm for $\vee$.

Although t-norms and t-conorms are studied in detail in Chapter 3, we only introduce two t-norms (t-conorms) here so we can show how fuzzy logic generalizes classical logic.

2.2 Propositional Logic

Logic is the analysis of methods of reasoning. The propositional logic is a logic which deals with propositions. A proposition is a sentence which is either true or false. The "true " and "false" are called the truth values. We denote these values by 1 and 0, respectively. Then to any sentence is assigned only 1 or 0. The propositional logic based on this preassumption is said to be the two-valued or classical propositional logic. Simple sentences or atomic propositions are denoted by $p, q, \ldots$ or $p_1, p_2, \ldots$.

Atomic propositions are combined to form more complicated propositions where the truth or falsity of the new sentences are determined by the truth or falsity of its component propositions.

Negation, denoted by $\sim$, is an operation on propositions. That is, if p is a proposition, then $\sim p$ is also a proposition, whose truth values are shown in the following truth table:

Table 2.1: Truth Table for Negation

p	$\sim p$
1	0
0	1

When p is true, $\sim p$ is false, when p is false, $\sim p$ is true.

Another common operation is conjunction "and". The conjunction of propositions p and q is denoted by $p \wedge q$ and has the following truth table:

Table 2.2: Truth Table for Conjunction

p	q	$p \wedge q$
1	1	1
1	0	0
0	1	0
0	0	0

$p \wedge q$ is true if and only if both p and q are true. p and q are called the conjuncts of $p \wedge q$. Note that there are cases (rows) in the table, corresponding to the number of possible assignments to truth values to p and q.

There is an operator on propositions corresponding to "or" called disjunction. The disjunction of proposition p and q is denoted by $p \vee q$. Its truth table is as follows:

Table 2.3: Truth Table for Disjunction

p	q	$p \vee q$
1	1	1
1	0	1
0	1	1
0	0	0

Thus, $p \vee q$ is false if and only if both p and q are false. The propositions p and q are said to be disjuncts.

Note that there is another "or" called "exclusive or", whose meaning is p or q but not both.

Another truth operation on propositions is called conditional: if p, then q. "If p, then q " is false when the antecedent p is true and the consequent q is false, otherwise it is true. We denote "If p, then q" by $p \rightarrow q$ and say that p implies q or that q is implied by p. Thus $\rightarrow$ has the following truth table:

Table 2.4: Truth Table for Implication

p	q	$p \rightarrow q$
1	1	1
1	0	0
0	1	1
0	0	1

Let us denote "p if and only if q" by $p \leftrightarrow q$. Such an expression is called a biconditional. Clearly $p \leftrightarrow q$ is true when and only when p and q have the same truth values. Two propositions that have the same truth values are said to be equivalent. Its truth table is given below:

Table 2.5: Truth Table for Equivalence

p	q	$p \leftrightarrow q$
1	1	1
1	0	0
0	1	0
0	0	1

The symbols $\sim, \wedge, \vee, \rightarrow, \leftrightarrow$ are called propositional connectives. Any proposition built up by application of these connectives has a truth value 1 or 0 which depends on the truth values of the constituent propositions. In order to make this dependence clear, we use the name well-defined formulas or simply formulas to an expression build up from the propositional symbols p, q, r, etc., by appropriate applications of the proposition connectives. Thus we define:

a. every propositional symbol is a formula;

Table 2.6: Truth table for $((p \wedge (\sim q)) \rightarrow r)$

p	q	r	$(\sim q)$	$(p \wedge (\sim q))$	$((p \wedge (\sim q)) \rightarrow r)$
1	1	1	0	0	1
1	1	0	0	0	1
1	0	1	1	1	1
1	0	0	1	1	0
0	1	1	0	0	1
0	1	0	0	0	1
0	0	1	1	0	1
0	0	0	1	0	1

b. if P and Q are formulas, then $(\sim P), (P \wedge Q), (P \vee Q), (P \rightarrow Q)$ and $(P \leftrightarrow Q)$ are formulas ; and

c. only those expressions are formulas that are determined by means of (a) and (b).

Example 2.2.1

The expressions p, $(\sim p_2)$, $(p_3 \wedge (\sim q))$, $(((\sim p) \vee q) \rightarrow r)$, $((\sim p) \leftrightarrow p) \leftrightarrow (q \rightarrow (r \vee s))$ are all formulas.

Corresponding to each assignment of truth values 1 or 0 to the propositional symbols occurring in a formula, there is a truth value for the formula based on the truth tables for the propositional connectives. Then every formula determines a truth function, which can be represented by a truth table. For example, the formula $((p \wedge (\sim q)) \rightarrow r)$ has the truth table given in Table 2.6.

We note that if there are n distinct symbols in a formula, then there are 2^n possible assignments of truth values to the propositional symbols, i.e., 2^n rows in the truth table.

A truth table of m rows defines a truth function of m components which in turn defines a function of m arguments, the arguments and values are the truth values 1 or 0. Therefore any formula determines a truth function.

A formula that is always true is called a tautology. A formula is a tautology if and only if its corresponding truth function takes only the value 1, or equivalently, if in its truth table, the column under the formula contains only 1's.

Example 2.2.2

The following formulas are tautologies:

1. $(p \vee (\sim p))$ (excluded middle law);
2. $(\sim (p \wedge (\sim p)))$, $(p \leftrightarrow (\sim (\sim p)))$, $((p \rightarrow q) \rightarrow p)$; and
3. $(p \rightarrow (p \vee q))$.

If P and Q are formulas and $(P \rightarrow Q)$ is a tautology, we say that P logically implies Q, or Q is a logical consequence of P. For example, $(p \wedge q)$ logically implies p, and $(\sim (\sim p))$ logically implies p.

If P and Q are formulas and $(P \leftrightarrow Q)$ is a tautology, we say that P and Q are logically equivalent. For example, p and $(\sim (\sim p))$ are equivalent, as are $(p \rightarrow q)$ and $((\sim p) \vee q)$.

A formula that is false for all possible truth values of its propositional symbols is called a contradiction. It's truth table has only 0's in the column under the formula. For example, $(p \wedge (\sim p))$ and $(p \leftrightarrow (\sim p))$ are contradictions.

We note that a formula P is a tautology if and only if $(\sim P)$ is a contradiction.

2.2.1 Exercises

1. Write the following propositions as formulas, using proposition symbols to stand for simple sentences.

   a. John is happy or it is raining.

   b. If Mr. Amin is successful, then Mrs. Amin is happy, and if Mrs. Amin is not happy, then Mr. Amin is not successful.

   c. Ali goes to the party and Joe does not, or Sam does not go to the party and Max goes to the theater.

   d. Maria goes to the movies if a comedy is playing.

   e. A sufficient condition for x to be odd is that x is prime.

   f. If x is positive, x^2 is positive.

2. Write the truth table for $((p \to q) \wedge p)$ and $((p \vee (\sim q)) \leftrightarrow q)$.

3. Determine whether the following are tautologies:

   a. $(p \to q) \to (p \vee q)$;

   b. $(((p \to q) \to q) \to q)$;

   c. $(p \to (q \to (q \to p)))$;

   d. $((q \to r) \to (p \to q)) \to (p \to q)$;

   e. $((p \vee (\sim (q \wedge r))) \to ((p \leftrightarrow q) \vee r))$.

4. Prove or disprove the following:

   a. $((\sim p) \vee q)$ is logically equivalent to $((\sim q) \vee p)$;

   b. $(p \leftrightarrow q)$ logically implies $(p \to q)$;

   c. p is logically implied by $(p \wedge q)$;

   d. $(p \vee q)$ is logically implied by $(p \wedge q)$;

   e. $(p \wedge q)$ is logically implied by $(p \to q)$;

   f. $((\sim q) \to (\sim p))$ is logically implied by $(p \leftrightarrow q)$;

   g. $(p \to q)$ is logically implied by $(p \vee q)$.

5. Determine whether each of the following is a tautology, a contradiction, or neither:

   a. $(p \to q) \leftrightarrow (\sim (p \wedge (\sim q)))$;

   b. $(\sim p) \to (p \wedge q)$;

   c. $p \wedge (\sim (p \vee q))$;

   d. $(p \to q) \leftrightarrow ((\sim p) \vee q)$;

   e. $(p \to q) \to ((q \to r) \to (p \to r))$.

6. Show that the following pairs are logically equivalent:

   a. $p \wedge (q \vee r)$ and $(p \wedge q) \vee (p \wedge r)$;

   b. $p \vee (q \wedge r)$ and $(p \vee q) \wedge (p \vee r)$;

   c. $\sim (p \vee q)$ and $(\sim p) \wedge (\sim q)$;

   d. $\sim (p \wedge q)$ and $(\sim p) \vee (\sim q)$.

7. If P is a formula involving only $\sim, \wedge$, and $\vee$ and P' arises from P by replacing each $\wedge$ by $\vee$ and each $\vee$ by $\wedge$, show that P is a tautology if and only if $(\sim P')$ is a tautology. Prove that if $P \to Q$ is a tautology, so is $Q' \to P'$, and if $P \leftrightarrow Q$ is a tautology, so is $P' \leftrightarrow Q'$.

2.3 Crisp Sets

By a crisp set, or a classical set, or simply a set we mean a collection of distinct well-defined objects. These objects are said to be elements or members of the set. We usually denote the sets by capital letters A, B, C, etc., and the members by a, b, c, etc. To denote a is an element of A we write $a \in A$. The negation of $a \in A$ is written $a \notin A$ and means that a does not belong to A. A set with no elements is called an empty set and will be denoted by ϕ.

We say that the set A is a subset of B written as $A \subseteq B$ if every element of A is also a member of B. We write $A = B$ if the sets A and B have the same elements. Therefore two sets A and B are equal if and only if $A \subseteq B$ and $B \subseteq A$. A set A is said to be a proper subset of B, written $A \subset B$ if $A \subseteq B$ but $A \neq B$. The set of all subsets of a given set A is called the power set of A and is denoted by $\mathcal{P}(A)$.

A set A with elements $a_1, a_2, \ldots, a_n$ is denoted by $A = \{a_1, a_2, \ldots, a_n\}$ and in this case we say that A is finite. There are several ways to denote sets describing their elements. For instance the set of even natural numbers is denoted by $E = \{2, 4, 6, \ldots\}$ or equivalently $E = \{2k | k \in \mathbf{N}\}$ where $\mathbf{N}$ is the set of natural numbers. A set A is infinite, or has an infinitely many elements, if it is not finite. A set is denumerable if it is in a one-to-one correspondence with the set of natural numbers. A set is countable if it is finite or denumerable.

When we are talking about the sets, it is assumed that all sets are subsets of a given set called universal set, usually denoted by X. Then a universal set is a set which contains all the possible elements we need for a particular discussion or application.

Let X be the universal set, we define operations on $\mathcal{P}(X)$ as follows. Let A and B be two sets (they are elements of $\mathcal{P}(X)$). Then by A^c, called the complement of A, we mean the set of all elements in X which are not members of A, or

$$A^c = \{a \in X | a \notin A\}. \tag{2.1}$$

The union $A \cup B$ of sets A and B is defined to be the set of all elements which are members of A or B or both, or

$$A \cup B = \{x \in U | x \in A \text{ or } x \in B\}. \tag{2.2}$$

The intersection$A \cap B$ of sets A and B is defined to be the set of all elements which are members of both A and B, in notation

$$A \cap B = \{x \in U | x \in A \text{ and } x \in B\}. \tag{2.3}$$

The fundamental properties of ^{c}, $\cup$, $\cap$, which are similar to $\sim$, $\vee$, $\wedge$, respectively, are:

$$\text{Idempotency} : A \cup A = A, A \cap A = A, \tag{2.4}$$

$$\text{Commutativity}: A \cup B = B \cup A, A \cap B = B \cap A,$$

$$\text{Associativity}: A \cup (B \cup C) = (A \cup B) \cup C,$$

$$\text{Associativity}: A \cap (B \cap C) = (A \cap B) \cap C, \tag{2.7}$$

$$\text{Absorption}: A \cup (A \cap B) = A, A \cap (A \cup B) = A, \tag{2.8}$$

$$\text{Distributivity}: A \cap (B \cup C) = (A \cap B) \cup (A \cap C), \tag{2.9}$$

$$\text{Distributivity}: A \cup (B \cap C) = (A \cup B) \cap (A \cup C), \tag{2.10}$$

$$\text{Identity}: A \cup \phi = A, A \cup X = X, A \cap X = A, A \cap \phi = \phi, \tag{2.11}$$

$$\text{Law of Contradiction}: A \cap A^c = \phi, \tag{2.12}$$

$$\text{Law of Excluded Middle}: A \cup A^c = X, \tag{2.13}$$

$$\text{Involution}: (A^c)^c = A, \tag{2.14}$$

$$\text{De Morgan law}: (A \cup B)^c = A^c \cap B^c, \tag{2.15}$$

$$\text{De Morgan law}: (A \cap B)^c = A^c \cup B^c. \tag{2.16}$$

We have a mathematical system, denoted by $(\mathcal{P}(X),^c, \cup, \cap, \phi, X)$, where X is the universal set, $\mathcal{P}(X)$ is the set of all subsets of X, ^{c} is complementation, $\cap$ is intersection, $\cup$ is union and ϕ is the empty set, which obeys all the laws given by equations (2.4) to (2.16). Now we turn our attention again to propositional logic.

If we consider the set of all formulas $\mathcal{F}$ of propositional logic defined in Section 2.1 and define the relation $\equiv$ on $\mathcal{F}$ as

$$P \equiv Q \quad \text{if and only if} \quad P \leftrightarrow Q, \tag{2.17}$$

then $\equiv$ defines an equivalence relation on $\mathcal{F}$.

A relation is an equivalence relation if it is reflexive, symmetric and transitive. Equivalence relations are studied in Section 7.4 of Chapter 7. Every equivalence relation produces equivalence classes. If P is a formula in $\mathcal{F}$, by the equivalence class $[P]$ we mean the set $[P] = \{Q \in \mathcal{F} | Q \leftrightarrow P\}$. We denote the set of all equivalence classes of $\mathcal{F}$, by $[\mathcal{F}/\equiv]$, i.e., $[\mathcal{F}/\equiv] = \{[P] | P \in \mathcal{F}\}$.

We define the operations $\sim, \vee, \wedge$ on $[\mathcal{F}/\equiv]$ by

$$\sim [P] = [\sim P], \tag{2.18}$$

$$[P] \vee [Q] = [P \vee Q]. \tag{2.19}$$

$$[P] \wedge [Q] = [P \wedge Q]. \tag{2.20}$$

If 1 stands for $(p \vee \sim p) \in \mathcal{F}$, a tautology, and 0 stands for $(p \wedge \sim p) \in \mathcal{F}$, a contradiction, we can define $[1] \in [\mathcal{F}/\equiv]$ and $[0] \in [\mathcal{F}/\equiv]$.

Now we have the mathematical system $([\mathcal{F}/\equiv], \sim, \vee, \wedge, [0], [1])$. We may identify this mathematical system with the previous one built from sets as follows: (1) identify $[\mathcal{F}/\equiv]$ with $\mathcal{P}(X)$; (2) $\sim$ with ^{c}; (3) $\vee$ with $\cup$; (4) $\wedge$

with $\cap$; (5) [0] with ϕ; and (6) [1] with X. Then equations (2.4)-(2.16) hold for $[P]$, $[Q]$, $[R]$ in $[\mathcal{F}/\equiv]$.

For example, consider equation (2.15). Substitute $[P]$ for A, $[Q]$ for B, $\sim$ for ^{c}, $\wedge$ for $\cap$ and $\vee$ for $\cup$. Then equation (2.15) becomes

$$\sim ([P] \vee [Q]) = (\sim [P]) \wedge (\sim [Q]). \tag{2.21}$$

In this way equations in crisp set theory become tautologies in propositional logic.

2.3.1 Exercises

To show that an equation $E_1 = E_2$ involving sets is true you show that E_1 is a subset of E_2 and you show that E_2 is a subset of E_1. To show that E_1 is a subset of E_2 you show that each element in the set E_1 also belongs to the set E_2. To show that $E_1 \neq E_2$ you find an element in E_1 (or E_2) that does not belong to E_2 (or E_1). It is usually easiest to work with finite sets (universal set X finite) when showing that $E_1 \neq E_2$.

1. Show that if A and B are two sets such that $A \subseteq B$, then $\mathcal{P}(A) \subseteq \mathcal{P}(B)$.

2. Show that If $A \subseteq B$ and $B \subseteq C$, then $A \subseteq C$.

3. Let $n(A)$ be the number of elements of the finite set A.

   a. Show that if $n(A) = k$, then $n(\mathcal{P}(A))=2^k$.

   b. Prove or disprove: If $A \subseteq B$, then $n(A) \leq n(B)$.

4. Prove or disprove the following:

   a. $A \cap (A^c \cap B) = \phi$;
   b. $(A \cup B) \cup B^c = X$;
   c. $A \cap (A^c \cup B) = A \cup B$;
   d. $(A^c \cup B) \cup (A \cup B^c) = X$;
   e. $(A \cap B^c) \cap (A^c \cap B) = X$;
   f. $(A \cup B^c) \cap (A \cup B) = B$.

5. Prove:

   a. $(A^c \cap B) \cup (A^c \cap B^c) = A^c$;
   b. $A \cup (A \cap B^c)^c = X$;
   c. $(A \cup X^c) \cup (A^c \cup B)^c = A$.

6. Prove or disprove the following:

   a. If $A \cap B = \phi$, then $A^c \cap B^c = \phi$;
   b. If $A \cap B = \phi, B \subseteq C$, then $A \cap C = \phi$.

7. Prove or disprove the following:

   a. $x \in \{\{x\}, \{x, y\}\}$;
   b. $\{x\} \in \{x\}$;
   c. $\phi = \{\phi\}$;
   d. $\phi \in \{\phi\}$;
   e. If $x \in A, A \in B$, then $x \in B$;

f. If $A \not\subset B, B \subset C$, then $A \not\subset C$;

g. If $A \subseteq B, x \notin B$, then $x \notin A$.

8. Give an example of a three element set A whose members are also subsets of A.

9. Prove that if $A \subseteq B$, then $A \cap B^c = \phi$.

10. Show that if M is a set, there is a one-to-one correspondence between $\mathcal{P}(M)$ and $\{0,1\}^M$ where A^B stands for the set of all functions from B to A. By a one-to-one correspondence we mean there is a function from $\mathcal{P}(M)$ onto $\{0,1\}^M$ which is a one-to-one function.

11. Substitute $[P]$ for A, $[Q]$ for B, $\sim$ for ^{c}, etc. as outlined in the text translating the following equations over into propositional logic, and then show, using the results in Section 2.2, that the equation you get in propositional logic is true. That is, show the two expressions are logically equivalent:

    a. Equation (2.6).

    b. Equation (2.9).

    c. Equation (2.12).

    d. Equation (2.15).

12. Another identification from propositional logic and set theory is to substitute $\rightarrow$ for $\subseteq$ in set theory to obtain correct expressions in propositional logic. Write down three correct equations in set theory using $\subseteq$, translate to propositional logic and then show that the resulting propositional logic statements are true.

2.4 Fuzzy Logic

The beginning of fuzzy logic is to allow truth values to be any number in the interval $[0, 1]$. If p is a atomic proposition, then we will now let $tv(p)$ denote the truth of p. So, $tv(p) \in [0, 1]$ for any proposition in fuzzy logic. $tv(p) = 1$ means that p is absolutely true, $tv(p) = 0$ is that p is absolutely false and $tv(p) = 0.65$ just means that the truth of p is 0.65. Fuzzy logic is an infinite valued logic in that truth values can range from zero to one. Like classical logic, fuzzy logic is concerned with the truth of propositions. However, in the real world propositions are often only partly true. It is hard to characterize the truth of "John is old" as unambiguously true or false if John is 60 years old. In some respects he is old, being eligible for senior citizen benefits at many establishments, but in other respects he is not old since he is not eligible for social security. So, in fuzzy logic we would allow tv(John is old) to take on other values in the interval $[0, 1]$ besides just zero and one.

What is added in fuzzy logic is that there are many ways to combine these truth values. Negation is the same with $tv(\sim p) = 1 - tv(p)$. To find $tv(p \wedge q)$, for two atomic propositions p and q, we usually employ a t-norm in fuzzy logic.

t-norms are studied in detail in Section 3.3 so for now we will introduce only two t-norms: (1) $T_m(a, b) = \min(a, b)$, for $a, b \in [0, 1]$; and (2) $T_b(a, b) = \max(0, a + b - 1)$, for $a, b \in [0, 1]$. T_m is called standard intersection (min) and T_b is called "bounded sum".

Therefore, in fuzzy logic we find the truth value of $p \wedge q$ as

$$tv(p \wedge q) = T(tv(p), tv(q)), \tag{2.22}$$

for t-norm T (T_m or T_b). This generalizes classical propositional logic (section 2.2) because if $tv(p)$ and $tv(q)$ only have the values zero or one, then we will get Table 2.2.

We obtain the truth value of $p \vee q$ in fuzzy logic by using a t-conorm, studied also in Section 3.3. The two t-conorms related to the two t-norms defined above are: (1) $C_m(a, b) = \max(a, b)$, for $a, b \in [0, 1]$, related to T_m; and (2) $C_b(a, b) = \min(1, a + b)$, for $a, b \in [0, 1]$, related to T_b. In fuzzy logic we compute the truth value of $p \vee q$ as

$$tv(p \vee q) = C(tv(p), tv(q)), \tag{2.23}$$

for t-conorm C (C_m or C_b). Also, if $tv(p)$ and $tv(q)$ only equal zero or one, then we get Table 2.3. So, in fuzzy logic no matter what t-norm and t-conorm we use, if the truth values are restricted to be only zero or one, fuzzy logic collapses back to classical logic.

Another basic operator of classical logic that we can fuzzify is implication in Table 2.4. Three different methods of translating implication to fuzzy logic are:

$$tv(p \to q) = tv(\sim p \vee q), \tag{2.24}$$

$$tv(p \to q) = \min(1, 1 - tv(p) + tv(q)), \tag{2.25}$$

and

$$tv(p \to q) = \max(1 - tv(p), \min(tv(p), tv(q))). \tag{2.26}$$

For example we would use a t-conorm to evaluate equation (2.24) as follows

$$tv(p \to q) = C(1 - tv(p), tv(q)), \tag{2.27}$$

for $C = C_m$ or C_b. In all cases, equations (2.24)-(2.26), we obtain Table 2.4 if we restrict the values of $tv(p)$ and $tv(q)$ to be only zero and one. However, these equations can produce different results when $tv(p)$ and $tv(q)$ can take on other values beside only zero and one.

Having fuzzified $\wedge$, $\vee$ and $\to$ we can go on to find the truth values of more complicated expressions. For example, in fuzzy logic we can compute the truth value of $(p \wedge q) \to r$ as

$$\min(1, 1 - \min(tv(p), tv(q)) + tv(r)), \tag{2.28}$$

using equation (2.25) for implication and T_m for $\wedge$.

Fuzzy logic can be extended in various ways. One extension allows truth values to be fuzzy numbers in $[0, 1]$. Fuzzy numbers are to be studied in Chapter 4. However, we will not pursue further development of fuzzy logic in this book.

2.4.1 Exercises

1. Show that equation (2.22) will give Table 2.2, if the truth values are only zero and one, and if the t-norm used is:

   a. T_m;

   b. T_b.

2. Show that equation (2.22) can give different values, using t-norms T_m and T_b, when the truth values can be any number in $[0,1]$.

3. Show that equation (2.23) will produce Table 2.3, if the truth values are only zero or one, and if the t-conorm is:

   a. C_m;

   b. C_b.

4. Show that equation (2.23) will have different values, using t-conorms C_m and C_b, when the truth values can be any number in $[0,1]$.

5. Show that equations (2.24)-(2.26) will all reduce to Table 2.4 when the truth values are only zero and one. In equation (2.24) use both C_m and C_b.

6. Show that equations (2.24)-(2.26) can all give different results when the truth values can be any number in $[0,1]$. Use both C_m and C_b in equation (2.24).

7. Find the truth value of $(p \wedge q) \rightarrow r$ using:

   a. equation (2.24) and either t-norm (t-conorm) for $\wedge$ ($\vee$);

   b. equation (2.26) and either t-norm for $\wedge$.

8. We saw in section 2.3, and the exercises in that section, how to obtain basic equations in propositional logic by translating equations (2.4)-(2.16) from set theory into propositional logic. But now, in fuzzy logic, not all of these basic equations of propositional logic remain true. Using T_m (or T_b) for $\wedge$, and using C_m (or C_b) for $\vee$ determine which of these basic equations are still true.

Chapter 3

Fuzzy Sets

3.1 Introduction

The basic concept of a fuzzy set is introduced in the next section. t-norms and t-conorms are used throughout fuzzy set theory and fuzzy logic and they are studied in the third section. t-norms (t-conorms) are used to compute the intersection (union) of fuzzy sets. Once we have intersection and union of fuzzy sets, we can study the algebra of fuzzy sets in section four. In Section 2.4 we notice that all the laws of crisp set theory (presented in Section 2.2) do not necessarily hold for fuzzy sets. Mixed fuzzy logic is introduced in Section 2.5 to show one method of getting fuzzy sets to obey all the basic laws of crisp set theory. α-cuts, or a way to represent a fuzzy set as a collection of nested crisp sets, is then discussed in section six. To initiate a calculus of continuous fuzzy subsets of the real numbers, determining the distance between these fuzzy sets comprises the final section of this chapter.

3.2 Fuzzy Sets

In order to see the connection between regular sets (now called crisp sets) and fuzzy sets we first review the basic material on crisp sets.

Let X be a universal set, which contains all the elements of interest for our present discussion(application). Let A be a subset of X. The characteristic function (or membership function) of A is a function on X, with values zero or one, so that it equals one at x in X whenever x is in A and otherwise it equals zero. We write its membership function as $A(x)$. Then $A(x) = 1$ if x is in A and $A(x) = 0$ if x does not belong to A. The characteristic function is sometimes written $\chi_A(x)$ but we will use $A(x)$ for this function.

The power set of X, written $\mathcal{P}$(X), is the set of all subsets of X. The subset of X having no elements is called the empty set ϕ. If A and B are two subsets of X then we say that A is a subset of B, $A \subseteq B$, whenever $A(x) = 1$

implies $B(x)$ is also equal to one. So A equals B if $A \subseteq B$ and $B \subseteq A$. The complement of A, A^c, is defined as $A^c(x) = 1 - A(x)$, for all x in X. Of course, $\phi^c = X$ and $X^c = \phi$.

To obtain our algebra of subsets of X we need to define intersection ($\cap$) and union ($\cup$).

C is the intersection of A and B, written $C = A \cap B$, if

$$C(x) = \begin{cases} 1, & \text{if} \quad A(x) = B(x) = 1 \\ 0, & \quad \text{otherwise.} \end{cases} \tag{3.1}$$

A and B are said to be disjoint whenever $A \cap B = \phi$, or $A(x)B(x) = 0$ for all x in X.

D is the union of A and B, $D = A \cup B$, if

$$D(x) = \begin{cases} 1, & \text{if} \quad A(x) = 1 \text{ } \mathit{or} \text{ } B(x) = 1 \\ 0, & \quad \text{otherwise.} \end{cases} \tag{3.2}$$

We have used the membership functions in all these definitions because that is what is used when we generalize crisp sets to fuzzy sets.

We now list the basic properties of complement, union and intersection. All sets are subsets of the same X.

$$\text{Involution}: (A^c)^c = A. \tag{3.3}$$

$$\text{Commutativity}: A \cup B = B \cup A, A \cap B = B \cap A. \tag{3.4}$$

$$\text{Associativity}: \quad (A \cup B) \cup C = A \cup (B \cup C), \tag{3.5}$$
$$(A \cap B) \cap C = A \cap (B \cap C). \tag{3.6}$$

$$\text{Distributivity}: \quad A \cap (B \cup C) = (A \cap B) \cup (A \cap C), \tag{3.7}$$
$$A \cup (B \cap C) = (A \cup B) \cap (A \cup C). \tag{3.8}$$

$$\text{Idempotency}: A \cap A = A, A \cup A = A. \tag{3.9}$$

$$\text{Law of Contradiction}: A \cap A^c = \phi. \tag{3.10}$$

$$\text{Law of Excluded Middle}: A \cup A^c = X. \tag{3.11}$$

$$\text{De Morgan}: \quad (A \cup B)^c = A^c \cap B^c, \tag{3.12}$$
$$(A \cap B)^c = A^c \cup B^c. \tag{3.13}$$

$$\text{Identity}: \quad A \cup \phi = A, A \cap \phi = \phi, \tag{3.14}$$
$$A \cup X = X, A \cap X = A. \tag{3.15}$$

$$\text{Absorbtion}: \quad A \cup (A \cap B) = A, \tag{3.16}$$

$$A \cap (A \cup B) = A. \tag{3.17}$$

Let us show how these may be proven using membership functions. Let us first show the De Morgan identity $(A \cap B)^c = A^c \cup B^c$ in equation (3.13). Let $C = (A \cap B)^c$ and $D = A^c \cup B^c$ and we show that $C(x) = D(x)$ for all x in X. Now $C(x) = 1$ if $(A \cap B)(x) = 0$ and $C(x) = 0$ for $(A \cap B)(x) = 1$. So we see

$$C(x) = \begin{cases} 1, & \text{if} \quad A(x) \textit{ or } B(x) = 0 \\ 0, & \text{if} \quad A(x) \textit{ and } B(x) = 1. \end{cases} \tag{3.18}$$

Also, $D(x) = 1$ if $A^c(x)$ or $B^c(x) = 1$ and $D(x) = 0$ for $A^c(x)$ and $B^c(x) = 0$. Hence

$$D(x) = \begin{cases} 1, & \text{if} \quad A(x) \textit{ or } B(x) = 0 \\ 0, & \text{if} \quad A(x) \textit{ and } B(x) = 1. \end{cases} \tag{3.19}$$

From equations (3.18) and (3.19), $C(x) = D(x)$, for all x in X, and this De Morgan law holds.

For another proof let us show the absorption law $A \cap (A \cup B) = A$ of equation (3.17). Let $C = A \cap (A \cup B)$. Then $C(x) = 1$ if $A(x) = 1$ and $(A \cup B)(x) = 1$ and $C(x) = 0$ if $A(x) = 0$ or $(A \cup B)(x) = 0$. Therefore $C(x) = 1$ if $A(x) = 1$ and $(A(x) = 1$ or $B(x) = 1)$ and $C(x) = 0$ if $A(x) = 0$ or $(A(x) = 0$ and $B(x) = 0)$. Hence, $C(x) = 1$ if $A(x) = 1$ and $C(x) = 0$ when $A(x) = 0$ and $C(x) = A(x)$ for all x.

We will also be using set products. If X and Y are two universal sets, then $X \times Y$ is the set of all ordered pairs (x, y) for x in X and y in Y. For finite sets X and Y we may list all of these ordered pairs. Let $X = \{x_1, x_2, x_3\}$ and $Y = \{y_1, y_2\}$. Then $X \times Y$ is $\{(x_1, y_1), (x_1, y_2), \ldots, (x_3, y_2)\}$ having six elements.

If R is a subset of $X \times Y$, then R is called a relation between X and Y, also called a relation on $X \times Y$. For example, if X is a set of men and Y is a set of women, then R could be "x is the husband of y". We let $\mathbf{R}$ denote the set of real numbers, and $\mathbf{R}^2 = \mathbf{R} \times \mathbf{R}$. A relation on $\mathbf{R}^2$ is "x is less than y". For this relation "x is less than y" we would have $R(x, y) = 1$ if $x < y$ and $R(x, y) = 0$ when $x \geq y$.

Before we introduce fuzzy sets let us say a few words about *sup* (supremum) versus max (maximum) and *inf*(infimum) versus min (minimum). Consider the set $S = \{x | 0 \leq x \leq 1\}$, or the unit interval $[0, 1]$. This set has a maximum member $x = 1$ so we would write $\max S = 1$. Now consider $S' = \{x | 0 \leq x < 1\}$ which has no maximum member so that $\max S'$ does not exist. In this case we would use *sup* and $supS' = 1$. The least upper bound of S' is called $supS'$. A number u is called an upper bound for S' if $x \leq u$ for all x in S' and $supS'$ is the smallest upper bound. Any subset of $\mathbf{R}$ that has an upper bound has a supremum. Therefore, "*sup*" is more general than max and $supS = \max S = 1$. So, in general, we will use "*sup*" even when the set can have a maximum value. The dual of *sup* is *inf* as the dual of max is min. For example if $S'' = \{x|0 < x \leq 1\}$, then $\min S = infS = 0$ but $infS'' = 0$, $\min S''$ does not exist.

Now we define fuzzy subsets of a universal set X. We will use the notation of placing a "bar" over a letter (symbol) to denote a fuzzy set. So $\overline{A}$, $\overline{B}$, ..., $\overline{\alpha}$, ... all represent fuzzy subsets of X. Fuzzy subsets are also defined by their membership function but now their values can be any number between zero and one. That is, $\overline{A}(x)$ the membership function for $\overline{A}$ a fuzzy subset of X, is any number in the interval $[0, 1]$. If $\overline{A}(x) = 1$ we say that x belongs to $\overline{A}$, $\overline{A}(x) = 0$ means that x does not belong to $\overline{A}$, and $\overline{A}(x) = 0.6$ says x has membership value 0.6 in $\overline{A}$. Crisp sets are special cases of fuzzy sets where $\overline{A}(x)$ is only zero or one. The universal fuzzy set is $\overline{X}(x) = 1$ for all x and the empty set is $\overline{\phi}(x) = 0$ for all x. The fuzzy power set of X, written $\mathcal{F}(X)$, is all fuzzy subsets of X.

Some examples of fuzzy sets are: (1) the set of young people; (2) the set of fast cars; (3) the set of smart math students. Chris, who is 25 years old, may have membership value 0.6 in the set of young people. Tina has membership value 1.0 in the collection of smart math students.

We say $\overline{A}$ is a fuzzy subset of $\overline{B}$, written $\overline{A} \leq \overline{B}$, if $\overline{A}(x) \leq \overline{B}(x)$ for all x. If $\overline{A} \leq \overline{B}$, then $\overline{A}(x) = 1$ implies $\overline{B}(x) = 1$. So, $\overline{A}$ equals $\overline{B}$ ($\overline{A} = \overline{B}$) if $\overline{A}(x) = \overline{B}(x)$ for all x in X.

The height of a fuzzy set $\overline{A}$ is defined as

$$ht(\overline{A}) = sup\{\overline{A}(x)|x \ in \ X\}. \tag{3.20}$$

A fuzzy set $\overline{A}$ is said to be normal whenever $ht(\overline{A}) = 1$.

The complement of $\overline{A}$, written $\overline{A}^c$, is defined as $\overline{A}^c(x) = 1 - \overline{A}(x)$. Hence, $\overline{X}^c = \overline{\phi}$, $\overline{\phi}^c = \overline{X}$.

To determine the algebra of fuzzy subsets of X we have to specify intersection and union. We start with intersection. Let $\overline{C} = \overline{A} \cap \overline{B}$. Clearly, $\overline{C}(x) = 1$ if $\overline{A}(x) = \overline{B}(x) = 1$ and $\overline{C}(x) = 0$ whenever $\overline{A}(x)$ or $\overline{B}(x) = 0$. But what is $\overline{C}(x)$ if $\overline{A}(x) = 0.7$ and $\overline{B}(x) = 0.4$? The value of $\overline{C}(x)$ will be a function of the two values $\overline{A}(x)$ and $\overline{B}(x)$. Let this unknown function be called $i(a, b)$, for intersection, with a, b in $[0, 1]$. Here we let $a = \overline{A}(x)$, $b = \overline{B}(x)$ so that $\overline{C}(x) = i(a, b)$.

This function $i(a, b)$ must have the properties: (1) $0 \leq a, b \leq 1$ and $i(a, b)$ is in $[0, 1]$; (2) $i(1, 1) = 1$; and (3) $i(0, 1) = i(1, 0) = i(0, 0) = 0$.

We write

$$\overline{C}(x) = i(\overline{A}(x), \overline{B}(x)), \tag{3.21}$$

for all x in X. Equation (3.21) defines the intersection of two fuzzy sets $\overline{A}$ and $\overline{B}$ by defining the membership function of $\overline{C}$.

Choices for $i(a, b)$ could be

$$i(a, b) = ab, \tag{3.22}$$

$$i(a, b) = \min(a, b), \tag{3.23}$$

and

$$i(a,b) = \sqrt{\max(0, a^2 + b^2 - 1)}. \tag{3.24}$$

Next let $\overline{D} = \overline{A} \cup \overline{B}$. We need $\overline{D} = 1$ if $\overline{A}(x)$ or $\overline{B}(x) = 1$ and $\overline{D}(x) = 0$ for $\overline{A}(x) = \overline{B}(x) = 0$. The function for union will be written $u(a,b)$ for a, b in $[0,1]$.

Its basic properties are: (1) $0 \leq a, b \leq 1$ and $u(a,b)$ is in $[0,1]$; (2) $u(1,1) = u(0,1) = u(1,0) = 1$; and (3) $u(0,0) = 0$. We have

$$\overline{D}(x) = u(\overline{A}(x), \overline{B}(x)), \tag{3.25}$$

for all x.

Choices for u are

$$u(a,b) = a + b - ab, \tag{3.26}$$

$$u(a,b) = \max(a,b), \tag{3.27}$$

and

$$u(a,b) = \min(1, \sqrt{a^2 + b^2}). \tag{3.28}$$

Given an $i(a,b)$ and $u(a,b)$ we can check to see if fuzzy sets will have all the basic properties of crisp sets given in equations (3.3)- (3.17). Let us look at the De Morgan law $(\overline{A} \cap \overline{B})^c = \overline{A}^c \cup \overline{B}^c$. If this is true, then we need to show

$$1 - i(\overline{A}(x), \overline{B}(x)) = u(1 - \overline{A}(x), 1 - \overline{B}(x)), \tag{3.29}$$

for all x. Or, letting $a = \overline{A}(x)$, $b = \overline{B}(x)$, then

$$1 - i(a,b) = u(1 - a, 1 - b), \tag{3.30}$$

for all a,b in$[0,1]$.

If this De Morgan law is not true, all you need to do is find one $\overline{A}$ and $\overline{B}$ (and X) so that equation (3.29) is false. Let us now look at two situations where we first show some property does not hold for fuzzy sets, and then show some other property does hold. These results all depend on the $i(a,b)$ and $u(a,b)$ used.

Consider the law of contradiction [equation(3.10)] $\overline{A} \cap \overline{A}^c = \overline{\phi}$ for fuzzy sets. Choose $i(a,b) = \min(a,b)$, $u(a,b) = \max(a,b)$, $X = \{small, medium, large\}$ with

$$\overline{A} = \{\frac{0.3}{small}, \frac{1.0}{medium}, \frac{0.6}{large}\}. \tag{3.31}$$

$\overline{A}$ is called a discrete fuzzy set and its membership function is given in equation (3.31) because $\overline{A}(small) = 0.3$, $\overline{A}(medium) = 1.0$ and $\overline{A}(large) = 0.6$. Then

$$\overline{A} \cap \overline{A}^c = \{\frac{0.3}{small}, \frac{0}{medium}, \frac{0.4}{large}\}, \tag{3.32}$$

which is not the same as $\overline{\phi}$. So, $\overline{A} \cap \overline{A}^c = \overline{\phi}$ may not hold using $i(a,b) = \min(a,b)$.

Using $i(a,b) = ab$, $u(a,b) = a+b-ab$ we may show that De Morgan's law $(\overline{A} \cup \overline{B})^c = \overline{A}^c \cap \overline{B}^c$ holds. First evaluate $(\overline{A} \cup \overline{B})^c$ to be $1-u(a,b)$, $a = \overline{A}(x)$, $b = \overline{B}(x)$, and $1 - u(a,b) = 1 - (a+b-ab)$. Next evaluate $\overline{A}^c \cap \overline{B}^c$ to be $i(1-a, 1-b) = (1-a)(1-b)$. We see that $1-(a+b-ab) = (1-a)(1-b)$ so that this De Morgan law is true for $i(a,b) = ab$, $u(a,b) = a+b-ab$.

A fuzzy subset of $X \times Y$ is called a fuzzy relation. For example let $X = \{John, Jim, Bill\}$, $Y = \{Fred, Mike, Sam\}$, and the fuzzy relation $\overline{R}$ between X and Y, which we will call "resemblance" might be shown as

	Fred	Mike	Sam
John	0.2	0.3	0.7
Jim	0.9	0.8	0
Bill	0.6	0.4	0.7

(3.33)

The table in equation (3.33) is called a type 1 fuzzy matrix. A type 1 fuzzy matrix has all its elements in $[0,1]$. The elements in the fuzzy matrix are all the membership values of the fuzzy relation. That is, $\overline{R}(John, Fred) = 0.2$, $\ldots$, $\overline{R}(Bill, Sam) = 0.7$.

There are different types of fuzzy sets. A discrete fuzzy set was already given in equation (3.31). In equation (3.31) X was finite but we can also have discrete fuzzy sets for X infinite. Let $X = \mathbf{R}$ and $\overline{A}$ a fuzzy subset of $\mathbf{R}$ so that $\overline{A}(x) \neq 0$ only for x in $\{1, 2, \ldots, 10\}$. Then we would write $\overline{A}$ as

$$\overline{A} = \{\frac{\mu_1}{1}, \frac{\mu_2}{2}, \ldots, \frac{\mu_3}{10}\}, \tag{3.34}$$

a discrete fuzzy set, where $\overline{A}(i) = \mu_i$, $1 \leq i \leq 10$.

As another example of a discrete fuzzy set is the diagnosis of psychiatric disorders

$$\overline{A} = \{\frac{\mu_1}{depression}, \frac{\mu_2}{schizophrenia}, \ldots\}. \tag{3.35}$$

At the other extreme of discrete fuzzy sets we have "continuous " fuzzy sets, fuzzy subsets of $\mathbf{R}$ whose membership functions are continuous. Figure 3.1 gives an example of a continuous fuzzy subset of $\mathbf{R}$. From Figure 3.1 we see $\overline{N}(0) = 0$, $\overline{N}(2) = 1$, $\overline{N}(3.5) = 0.6351$, $\overline{M}(2) = 0.5$, $\overline{M}(8) = 0$, etc.

The fuzzy sets we have been describing are all called regular fuzzy sets, or just fuzzy sets for short. A type 2 fuzzy set has fuzzy membership values. If $X = \{x_1, \ldots, x_n\}$ and $\overline{A}(x_i) = \overline{\mu_i}$, $1 \leq i \leq n$, where $\overline{\mu_i}$ is a fuzzy subset of $[0,1]$, then $\overline{A}$ is a type 2 discrete fuzzy set. The membership values are not known exactly and are described as fuzzy sets. A $\overline{\mu_i}$ is shown in Figure 3.2 meaning approximately 0.5.

A level 2 fuzzy set has fuzzy members. For example

$$\overline{A} = \{\frac{0.3}{\overline{x_1}}, \frac{1.0}{\overline{x_2}}, \frac{0.7}{\overline{x_3}}\}, \tag{3.36}$$

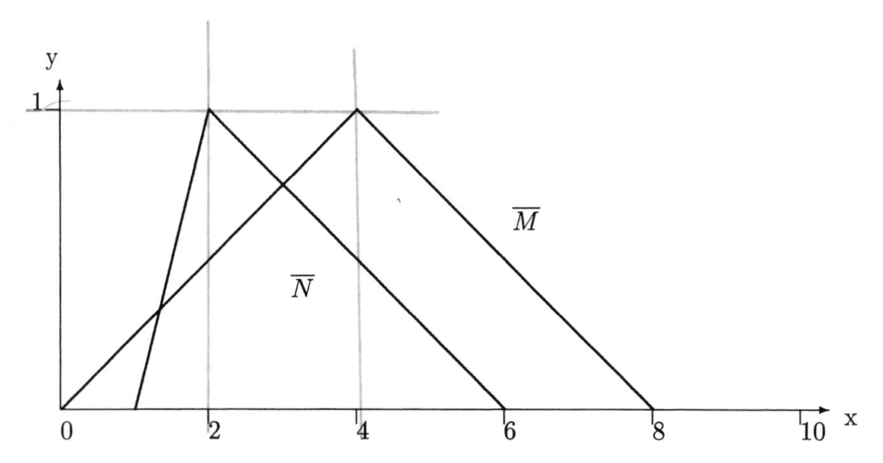

Figure 3.1: Continuous Fuzzy Set $\overline{N}$

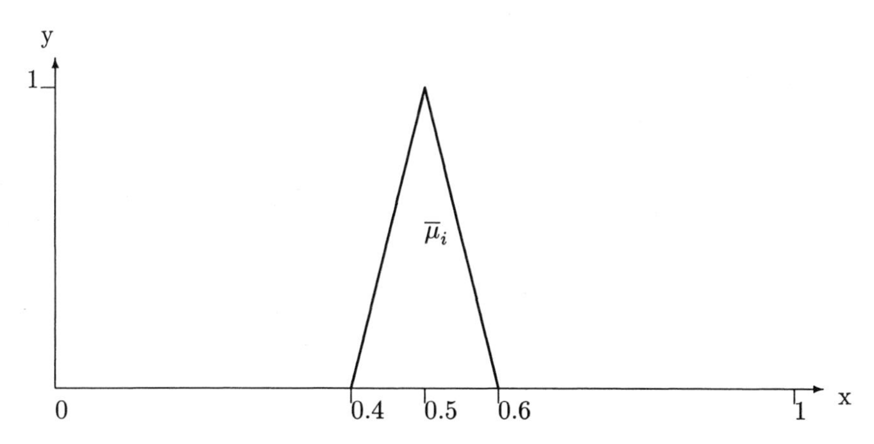

Figure 3.2: Approximately 0.5

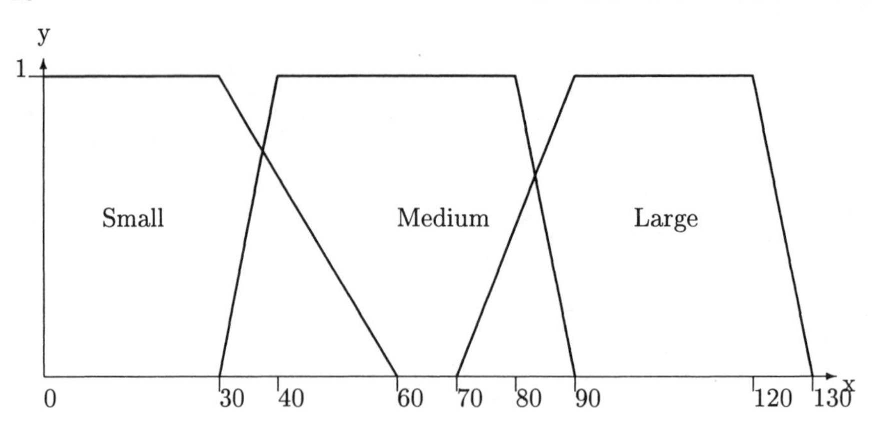

Figure 3.3: Fuzzy Sets for Level 2 Fuzzy Set

is a level 2 discrete fuzzy set. An example of this type of fuzzy set is given by the discrete fuzzy set in equation (3.31) where small, medium and large are defined by the fuzzy sets in Figure 3.3. Level 2 fuzzy sets come from $\mathcal{F}(\mathcal{F}(X))$, the fuzzy power set of $\mathcal{F}(X)$.

Traditional mathematics is based on crisp set theory. Fuzzy mathematics is based on fuzzy set theory. Traditional mathematics becomes a special case of fuzzy mathematics when we restrict all membership functions to have values only zero and one.

3.2.1 Exercises

1. Show, using membership functions, that the following are true for crisp sets:

   a. $(A \cup B)^c = A^c \cap B^c$.
   b. $A \cap (B \cup C) = (A \cap B) \cup (A \cap C)$.
   c. $A \cup (A \cap B) = A$.
   d. $A \subseteq B$ if and only if $B^c \subseteq A^c$.

2. Using $i(a,b) = \min(a,b)$, $u(a,b) = \max(a,b)$ determine if the following are true for fuzzy sets:

   a. $\overline{A} \cap (\overline{B} \cup \overline{C}) = (\overline{A} \cap \overline{B}) \cup (\overline{A} \cap \overline{C})$.
   b. $(\overline{A} \cap \overline{B})^c = (\overline{A}^c \cup \overline{B}^c)$.
   c. $\overline{A} \cap (\overline{A} \cup \overline{B}) = \overline{A}$.
   d. $\overline{A} \cup \overline{A}^c = \overline{X}$, $\overline{A} \cap \overline{A}^c = \overline{\phi}$.

3. Using $i = (a,b) = ab, u(a,b) = a + b - ab$ determine if the following are true for fuzzy sets:

   a. $\overline{A} \cap \overline{A}^c = \overline{\phi}$.
   b. $\overline{A} \cup (\overline{B} \cap \overline{C}) = (\overline{A} \cup \overline{B}) \cap (\overline{A} \cup \overline{C})$.
   c. $\overline{A} \cup (\overline{A} \cap \overline{B}) = \overline{A}$.
   d. $\overline{A} \subseteq \overline{B}$ implies $\overline{A} \cup \overline{B} = \overline{B}$.

4. Can you find a $i(a,b)$ and $u(a,b)$, subject to $i(1,1) = 1, \ldots, u(0,0) = 0$ given in the text, so that equations (3.3)-(3.17) are all true for fuzzy sets?

5. Discuss intersection and union for type 2 fuzzy sets.

6. Discuss intersection and union for level 2 fuzzy sets.

7. If $\overline{A} \leq \overline{B}$ and $\overline{B} \leq \overline{C}$, then does $\overline{A} \leq \overline{C}$ hold?

8. Answer the following questions:

   a. When does $X \times Y = Y \times X$?
   b. If A is crisp subset of X and B,C are crisp subset of Y, does $A \times (B \cap C)$ equal $(A \times B) \cap (A \times C)$?

9. Using $i(a,b) = \min$, show

$$i(\overline{A}(x), \overline{B}(x)) \leq ht(\overline{A} \cap \overline{B}) \leq \max\{ht(\overline{A}), ht(\overline{B})\},$$

   for all x.

10. Discuss the elements in the sets:

    a. $\mathcal{P}(\mathcal{P}\ (X))$.

    b. $\mathcal{P}(\mathcal{F}\ (X))$.

    c. $\mathcal{F}(\mathcal{P}\ (X))$.

    $\mathcal{F}(\mathcal{F}\ (X))$ gives level 2 fuzzy sets. Do any of the above (a, b, c) give type 2 fuzzy sets?

11. Give other examples of fuzzy sets besides "young", "fast cars", "smart", ...given in the text.

12. Can we use the following functions for $i(a, b)$?

    a.

$$(ab)/\max\{a, b, 0.5\}.$$

    b.

$$\max\{0, \frac{a+b+ab-1}{2}\}.$$

    c.

$$1 - \sqrt{(1-a)^2 + (1-b)^2 - (1-a)^2(1-b)^2}.$$

13. Can we use the following functions for $u(a, b)$?

    a.

$$\sqrt{a^2 + b^2 - a^2b^2}.$$

    b.

$$\min\{1, a + b + 2ab\}.$$

    c.

$$1 - \{\max\{0, \sqrt{(1-a)^2 + (1-b)^2 - 1}\}\}.$$

14. Let

$$\overline{A} = \{\frac{\overline{\mu}_1}{\overline{x}_1}, \cdots, \frac{\overline{\mu}_n}{\overline{x}_n}\},$$

where the membership value $\overline{\mu}_i$ is fuzzy and the members $\overline{x}_i$ are also fuzzy. Present some real world applications of such fuzzy sets (type 2 - level 2 fuzzy sets).

3.3 t-norms, t-conorms

The functions used for intersection of fuzzy sets ($i(a,b)$) are called t-norms and those used for union ($u(a,b)$) are called t-conorms. We will study t-norms first.

A t-norm T is a function $z = T(a,b)$, $0 \leq a, b, z \leq 1$, having the following four properties:

1. $T(a,1) = a$;
2. $T(a,b) = T(b,a)$;
3. if $b_1 \leq b_2$, then $T(a,b_1) \leq T(a,b_2)$;
4. $T(a,T(b,c)) = T(T(a,b),c)$.

Property 1 is a boundary condition implying $T(1,1) = 1$, $T(0,1) = 0$. Property 2 then says that $T(1,0) = 0$ too. From Property 3 we see that $0 \leq 1$ implies $T(0,0) \leq T(0,1) = 0$ and $T(0,0) = 0$. T has those three properties required of an intersection $i(a,b)$ given in the previous Section 3.2. In addition T is symmetric (property 2) and non-decreasing in both arguments (property 2 and 3). Also it is associative (Property 4) which we will need later in this section.

If $\overline{A}$ and $\overline{B}$ are fuzzy subsets of X and $\overline{C} = \overline{A} \cap \overline{B}$, then

$$\overline{C}(x) = T(\overline{A}(x), \overline{B}(x)) \tag{3.37}$$

for some t-norm T.

The basic t-norms are

$$T_m(a,b) = \min(a,b) \tag{3.38}$$

$$T_b(a,b) = \max(0, a+b-1) \tag{3.39}$$

$$T_p(a,b) = ab \tag{3.40}$$

$$T^*(a,b) = \begin{cases} a, & \text{if} \quad b=1 \\ b, & \text{if} \quad a=1 \\ 0, & \quad \text{otherwise.} \end{cases} \tag{3.41}$$

T_m is called standard intersection, T_b is bounded sum, T_p is algebraic product and T^* is drastic intersection.

It is not too difficult to see (see the problems at the end of this section) that

$$T^*(a,b) \leq T_b(a,b) \leq T_p(a,b) \leq T_m(a,b), \tag{3.42}$$

for all a, b in $[0,1]$.

In fact, if T is any t-norm, then

$$T^*(a,b) \leq T(a,b) \leq T_m(a,b), \tag{3.43}$$

for all a, b in $[0, 1]$.

A t-conorm C is a function $z = C(a, b)$, $0 \leq a, b, z \leq 1$, having the following four properties:

1. $C(a, 0) = a$;
2. $C(a, b) = C(b, a)$;
3. if $b_1 \leq b_2$, then $C(a, b_1) \leq C(a, b_2)$;
4. $C(a, C(b, c)) = C(C(a, b), c)$.

Properties 1,2 and 3 give $C(1,1) = C(0,1) = C(1,0) = 1$, $C(0,0) = 0$ the basic properties of the union function $u(a, b)$ in Section 3.2.

If $\overline{A}$ and $\overline{B}$ are two fuzzy subsets of X and $\overline{D} = \overline{B} \cup \overline{B}$, then

$$\overline{D}(x) = C(\overline{A}(x), \overline{B}(x)), \tag{3.44}$$

for all x in X, and some t-conorm C.

The basic t-conorms are

$$C_m(a, b) = \max(a, b), \tag{3.45}$$

$$C_b(a, b) = \min(1, a + b), \tag{3.46}$$

$$C_p(a, b) = a + b - ab, \tag{3.47}$$

$$C^*(a, b) = \begin{cases} a, & \text{if} \quad b = 0 \\ b, & \text{if} \quad a = 0 \\ 1, & \text{otherwise.} \end{cases} \tag{3.48}$$

C_m is standard union, C_b is bounded sum, C_p is algebraic sum, and C^* is drastic union.

We see (see the problems) that

$$C_m(a, b) \leq C_p(a, b) \leq C_b(a, b) \leq C^*(a, b), \tag{3.49}$$

for all a, b in $[0, 1]$, and if C is any t-conorm, then

$$C_m(A, b) \leq C(a, b) \leq C^*(a, b). \tag{3.50}$$

In practice one usually uses a pair of T and C which are dual. We say T and C are dual when

$$T(a, b) = 1 - C(1 - a, 1 - b), \tag{3.51}$$

$$C(a, b) = 1 - T(1 - a, 1 - b). \tag{3.52}$$

The following are dual: (1) T_m and C_m; (2) T_b and C_b; (3) T_p and C_p; (4) T^* and C^*,

It is interesting to compare the results using these different pairs of dual t-norms and t-conorms. Apply T_m, C_m and T^*, C^* to the continuous fuzzy

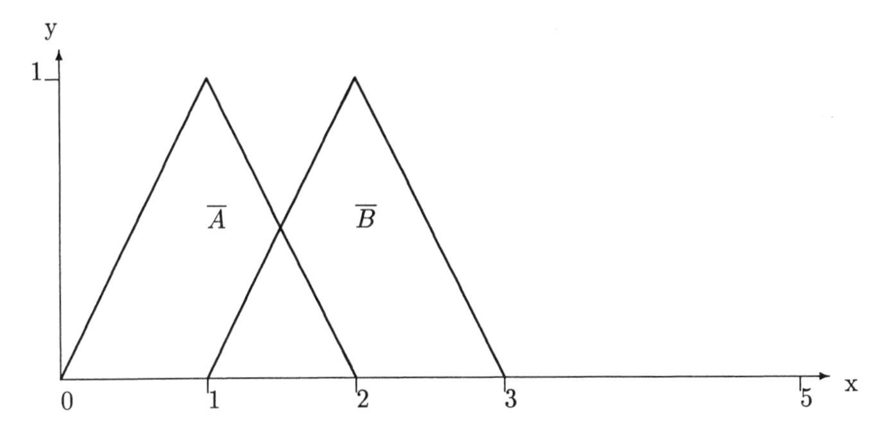

Figure 3.4: Continuous Fuzzy Sets $\overline{A}$ and $\overline{B}$

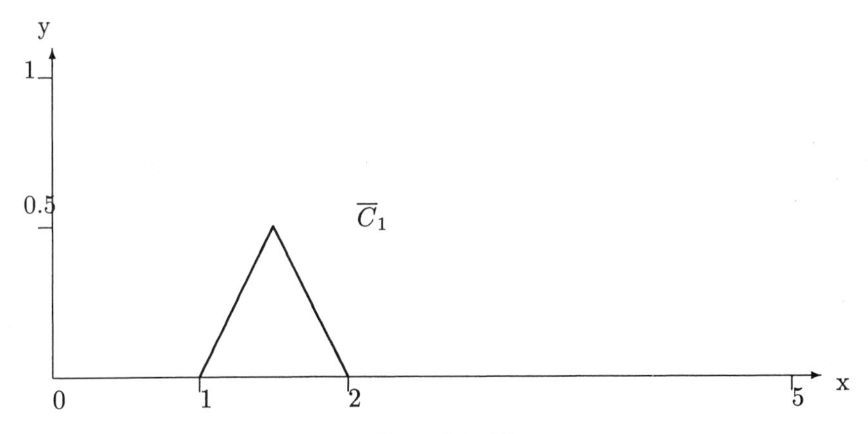

Figure 3.5: $\overline{C}_1 = \overline{A} \cap \overline{B}$ Using T_m

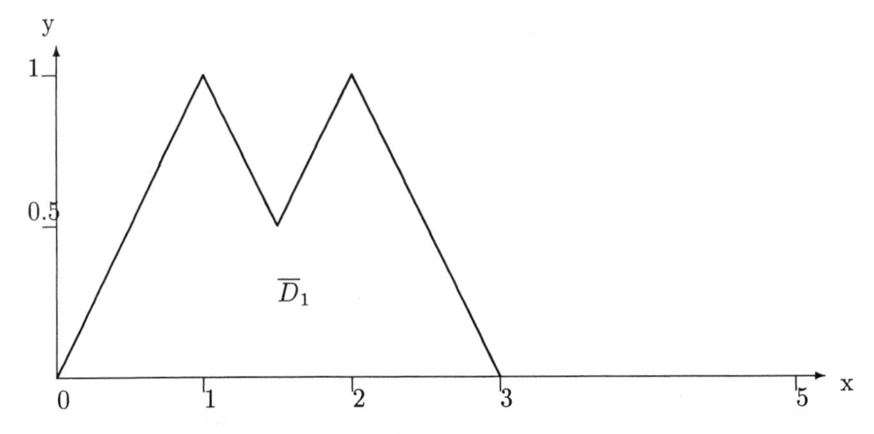

Figure 3.6: $\overline{D}_1 = \overline{A} \cup \overline{B}$ Using C_m

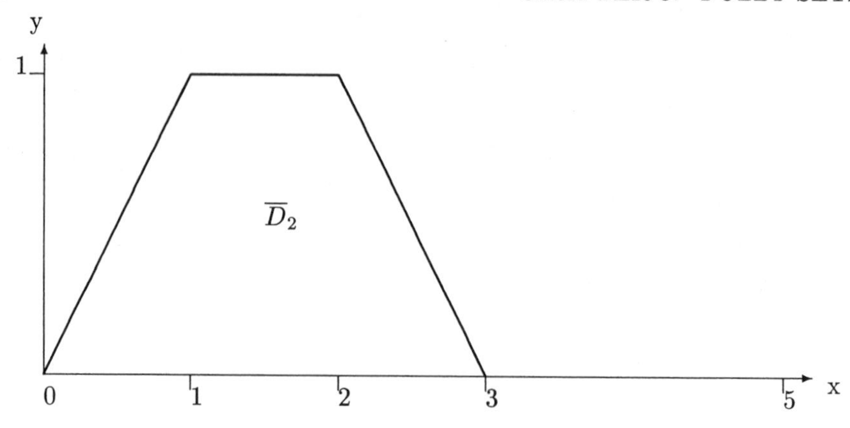

Figure 3.7: $\overline{D}_2 = \overline{A} \cup \overline{B}$ Using C^*

sets in Figure 3.4. Using T_m we get $\overline{C_1}$ in Figure 3.5 and $\overline{D_1}$ from C_m in Figure 3.6. If $\overline{C_2} = \overline{A} \cap \overline{B}$ using T^*, then $\overline{C_2} = \overline{\phi}$. Using C^* to get $\overline{D_2} = \overline{A} \cup \overline{B}$ we obtain the result in Figure 3.7.

Now we use the associativity property of T and C to extend them to n arguments. That is, we wish to define $T(a_1, a_2, \ldots, a_n)$ and $C(a_1, a_2, \ldots, a_n)$ for a_i in $[0,1]$, $1 \le i \le n$.

For T_m and C_m we easily obtain:

$$T_m(a_1, \ldots, a_n) = \min(a_1, \ldots, a_n), \tag{3.53}$$

$$C_m(a_1, \ldots, a_n) = \max(a_1, \ldots, a_n). \tag{3.54}$$

For T_b and C_b we obtain:

$$T_b(a_1, \ldots, a_n) = \max(0, \sum_{i=1}^{n} a_i - n + 1), \tag{3.55}$$

$$C_b(a_1, \ldots, a_n) = \min(1, \sum_{i=1}^{n} a_i). \tag{3.56}$$

Let us see how we might establish equations (3.55) and (3.56). We start with $T_b(a_1, a_2, a_3) = T_b(a_1, T(a_2, a_3))$ which is

$$\max(0, a_1 + \max(0, a_2 + a_3 - 1) - 1). \tag{3.57}$$

If $\max(0, a_2 + a_3 - 1) = a_2 + a_3 - 1$, then equation (3.57) is

$$\max(0, a_1 + a_2 + a_3 - 2) \tag{3.58}$$

that is the same as (3.55) for $n = 3$.

Now suppose that $\max(0, a_2 + a_3 - 1) = 0$ or $a_2 + a_3 - 1 < 0$. Then equation (3.57) is

$$\max(0, a_1 - 1) \tag{3.59}$$

which is zero since $a_1 \leq 1$. But equation (3.55), for $n = 3$, is also zero since $a_1 + a_2 + a_3 - 2 < 0$ when $a_2 + a_3 - 1 < 0$.

Next we look at $C_b(a_1, a_2, a_3) = C_b(a_1, C_b(a_2, a_3))$ which equals

$$\min(1, a_1 + \min(1, a_2 + a_3)). \tag{3.60}$$

We show that this equals

$$\min(1, a_1 + a_2 + a_3). \tag{3.61}$$

In equation (3.60) assume $a_2 + a_3 \leq 1$ so that $\min(1, a_2 + a_3) = a_2 + a_3$ and then it equals (3.61). So let $a_2 + a_3 > 1$. Then the min equals 1 and both equation (3.60) and (3.61) equal one.

For T_p we easily see

$$T_p(a_1, \ldots, a_n) = a_1 a_2 \ldots a_n, \tag{3.62}$$

but the formula for C_p is more complicated. When $n = 3$ we see

$$C_p(a_1, a_2, a_3) = a_1 + a_2 + a_3 - a_1 a_2 - a_1 a_3 - a_2 a_3 + a_1 a_2 a_3. \tag{3.63}$$

3.3.1 Exercises

1. Show:

   a. $T^*(a,b) \leq T_b(a,b)$, for all a,b in $[0,1]$.

   b. $T_b(a,b) \leq T_p(a,b)$, for all a,b in $[0,1]$.

   c. $T_p(a,b) \leq T_m(a,b)$, for all a,b in $[0,1]$.

2. Show that for any t-norm T:

   a. $T(a,b) \leq T_m(a,b)$, for all a,b in $[0,1]$.

   b. $T^*(a,b) \leq T(a,b)$, for all a,b in $[0,1]$.

3. Show:

   a. $C_m(a,b) \leq C_p(a,b)$, for all a,b in $[0,1]$.

   b. $C_p(a,b) \leq C_b(a,b)$, for all a,b in $[0,1]$.

   c. $C_b(a,b) \leq C^*(a,b)$, for all a,b in $[0,1]$.

4. Show, for any t-conorm C:

   a. $C(a,b) \leq C^*(a,b)$, for all a,b in $[0,1]$.

   b. $C_m(a,b) \leq C(a,b)$, for all a,b in $[0,1]$.

5. Using $\overline{A}$ and $\overline{B}$ in Figure 3.4 draw pictures, as in Figures 3.5-3.7 of $\overline{A} \cap \overline{B}$ and $\overline{A} \cup \overline{B}$ using:

   a. T_p and C_p.

   b. T_b and C_b.

6, Using $\overline{B}$ of Figure 3.4 draw pictures of $\overline{B} \cup \overline{B}^c$:

   a. Using T_m and C_m.

   b. Using T_b and C_b.

   c. Discuss the differences in the result of a. and b. That is, does either method produce $\overline{X}$?

7. Find the correct expression for $C_p(a_1, a_2, \ldots, a_n)$ where:

   a. n=4.

   b. n=5.

8. Find formulas for:

   a. $T^*(a_1, a_2, \ldots, a_n)$.

   b. $C^*(a_1, a_2, \ldots, a_n)$.

9. Show that:

   a. C_p and T_p are dual.

   b. C^* and T^* are dual.

10. Let T and C be dual t-norms and t-conorms, respectively. Show that the De Morgan laws [equations (3.12)-(3.13)] must hold.

11. Do any of the t-norms T_m, T_p, T_b or T^* have the property $T(a, a) = a$, a in $[0, 1]$, or they are idempotent?

12. Do any of the t-conorms C_m, C_p, C_b or C^* have the property $C(a, a) = a$ (idempotent), a in $[0, 1]$?

13. Show that C_m and C^* give the smallest and largest, respectively, union of two fuzzy sets. Or, let $\overline{D_1} = \overline{A} \cup \overline{B}$ using C_m, $\overline{D_2} = \overline{A} \cup \overline{B}$ using C^*, $\overline{D} = \overline{A} \cup \overline{B}$ using any t-norm, then $\overline{D_1}$ is a fuzzy subset of $\overline{D}$ and $\overline{D}$ is a fuzzy subset of $\overline{D_2}$. That is, show $\overline{D_1} \leq \overline{D} \leq \overline{D_2}$.

14. Using the notation of problem 13, show that T_m and T^* give the largest and smallest, respectively, intersection of two fuzzy sets.

15. Show that equation (2.22) will produce Table 2.2 for any t-norm T when the truth values are only zero and one.

16. Show that equation (2.23) will give Table 2.3 for any t-conorm C if the truth values are zero or one.

17. Let:

$$tv(p \to q) = \sup\{x \in [0, 1] | T(tv(p), x) \leq tv(q)\},$$

for any t-norm T. Show that this equation will compute Table 2.4 when the truth values are 0 or 1.

3.4 Algebra of Fuzzy Sets

Now we would like to evaluate and simplify expressions involving fuzzy sets $\overline{A}$, $\overline{B}$, $\overline{C}$, ... and intersections, unions and complements. However we must be careful because certain formulas involving crisp sets [like equations (3.3)–(3.17)] may not be true for fuzzy sets. So the first thing to do is to check and see which of these formulas are true, and which can be false, when using fuzzy sets. We will only be using the dual pairs T_m and C_m, T_p and C_p, T_b and C_b, and T^* and C^*. All fuzzy sets are fuzzy subsets of universal set X.

We start with $(\overline{A}^c)^c = \overline{A}$ from equation (3.3). But this is clearly always true since $(\overline{A})^c(x) = 1 - \overline{A}(x)$ so that $(\overline{A}^c)^c(x) = 1 - (1 - \overline{A}(x)) = \overline{A}(x)$ all x in X. Next let us look at the commutativity property in equation (3.4). Let $\overline{A}(x) = a$, $\overline{B}(x) = b$ for a, b in $[0,1]$. Now $(\overline{A} \cup \overline{B})(x) = C(\overline{A}(x), \overline{B}(x)) = C(a,b)$ and $(\overline{B} \cup \overline{A})(x) = C(\overline{B}(x), \overline{A}(x)) = C(b,a)$. But $C(a,b) = C(b,a)$ for all t-conorms C so that $\overline{A} \cup \overline{B} = \overline{B} \cup \overline{A}$. Similarly, $T(a,b) = T(\overline{A}(x), \overline{B}(x)) = T(b,a) = T(\overline{B}(x), \overline{A}(x)) = (\overline{A} \cap \overline{B})(x) = (\overline{B} \cap \overline{A})(x)$ and $\overline{A} \cap \overline{B} = \overline{B} \cap \overline{A}$. The commutativity property holds for fuzzy sets.

The associativity formulas [equations (3.5) and (3.6)] are also true for fuzzy sets because in terms of t-norms and t-conorms they are

$$C(C(a,b),c) = C(a, C(b,c)), \tag{3.64}$$
$$T(T(a,b),c) = T(a, T(b,c)), \tag{3.65}$$

where $\overline{C}(x) = c$ in $[0,1]$. Equations (3.64) and (3.65) are true for all t-conorms and t-norms.

The distributive laws, equations (3.7) and (3.8), are

$$T(a, C(b,c)) = C(T(a,b), T(a,c)), \tag{3.66}$$
$$C(a, T(b,c)) = T(C(a,b), C(a,c)), \tag{3.67}$$

for all a, b, c in$[0,1]$. It is not clear if equations (3.66) and (3.67) are true for $0 \leq a, b, c \leq 1$ so they need to be checked for our dual pairs.

The idempotent law states that $\overline{A} \cap \overline{A} = \overline{A}$ and $\overline{A} \cup \overline{A} = \overline{A}$. For this to be true we need

$$T(a,a) = a, \tag{3.68}$$
$$C(a,a) = a, \tag{3.69}$$

for all a in $[0,1]$.

Equations (3.68) and (3.69) say that T and C have to be idempotent. The only idempotent t-norm is T_m and the only idempotent t-conorm is C_m (see the exercises). So the idempotent laws hold for fuzzy sets only if we use T_m and C_m.

Next we come to the law of contradiction $\overline{A} \cap \overline{A}^c = \overline{\phi}$ and the law of excluded middle $\overline{A} \cup \overline{A}^c = \overline{X}$. We saw in Section 3.2 that $\overline{A} \cap \overline{A}^c \neq \phi$ using T_m (equation (3.32)). Using the same example (equation (3.31)) we easily see using C_m that $\overline{A} \cup \overline{A}^c \neq \overline{X}$. So these two laws may not hold if we use T_m and C_m. We need to investigate if these laws are true or false if we employ T_p and C_p, T_b and C_b, or T^* and C^*.

Now the De Morgan laws (equations (3.12) and (3.13)) hold for T and C dual (Exercise 10 in Section 3.3.1). Hence, the De Morgan formulas are true for T_m and C_m, or T_p and C_p, or T_b and C_b, or T^* and C^*.

The identities (equations (3.14) and (3.15)) are all true for all t-norms and t-conorms. The formula $\overline{A} \cup \overline{\phi} = \overline{A}$ translates to $C(a, 0) = a$ which is true and $\overline{A} \cap \overline{X} = \overline{A}$ is $T(a, 1) = a$. Also, $\overline{A} \cap \overline{\phi}$ is $T(a, 0) = 0$ (see the exercises) and $\overline{A} \cup \overline{X} = \overline{X}$ is $C(a, 1) = 1$ (see the exercises).

Finally we get to the absorbtion laws in equations (3.16) and (3.17). The expression $\overline{A} \cup (\overline{A} \cap \overline{B}) = \overline{A}$ is

$$C(a, T(a, b)) = a, \tag{3.70}$$

and $\overline{A} \cap (\overline{A} \cup \overline{B}) = \overline{A}$ is

$$T(a, C(a, b)) = a, \tag{3.71}$$

for all a, b in $[0, 1]$. Equations (3.70) and (3.71) must be checked to see if they are true or false.

In summary, this is what needs to be done: (1) see if the distributive laws hold; (2) check the law of contradiction and the law of the excluded middle for only T_p , C_p and T_b, C_b and T^*, C^*; (3) see if the absorbtion laws are true.

3.4.1 Exercises

To show a certain formula is true, write in terms of t-norms and t-conorms, then show the resulting equation is valid for all a, b (and c) in $[0, 1]$. To show a formula is not true you need only one example and usually the simplest one involves discrete fuzzy sets from a universal set having only three members.

1. Determine if the distributive laws are valid using:

   a. T_m and C_m.

   b. T_p and C_p.

   c. T_b and C_b.

   d. T^* and C^*.

2. Show that the only idempotent t-norm is T_m.

3. Show that the only idempotent t-conorm is C_m.

4. Determine if the law of contradiction holds for:

   a. T_p and C_p.

   b. T_b and C_b.

   c. T^* and C^*.

5. Determine if the law of the excluded middle holds for:

   a. T_p and C_p.

   b. T_b and C_b.

   c. T^* and C^*.

6. Show, using the definition of a t-norm, that $T(a, 0) = 0$ for all a in $[0, 1]$.

7. Show, using the definition of a t-conorm that $C(a, 1) = 1$ for all a in $[0, 1]$.

8. Determine if $\overline{A} \cup (\overline{A} \cap \overline{B}) = \overline{A}$ is true for:

   a. T_m and C_m.

   b. T_p and C_p.

   c. T_b and C_b.

   d. T^* and C^*.

9. Determine if $\overline{A} \cap (\overline{A} \cup \overline{B}) = \overline{A}$ holds if we use:

   a. T_m and C_m.

   b. T_p and C_p.

c. T_b and C_b.

d. T^* and C^*.

10. List all the basic laws of crisp set theory (equations (3.3)–(3.17)) that must hold for fuzzy sets when we use:

    a. T_m and C_m.

    b. T_p and C_p.

    c. T_b and C_b.

    d. T^* and C^*.

3.5 Mixed Fuzzy Logic

Most people use T_m and C_m in all their calculations with fuzzy sets. However, you need not always use the same t-norm and t-conorm. Mixed fuzzy logic is when you use $T = T_1$, $C = C_1$, usually dual operators, for certain calculations and switch to $T = T_2$, $C = C_2$, probably also dual operators, for other calculations. We will look at two applications of mixed fuzzy logic in this section but there can be many more since we will be using t-norms and t-conorms throughout the book.

We found in the previous section that using T_m, C_m, or T_p, C_p, or T_b, C_b, or T^*, C^* all the basic laws of crisp sets (equations (3.3)–(3.17)) do not hold for fuzzy sets. But equations (3.3)–(3.17) can be true for fuzzy sets if we use mixed fuzzy logic.

Let Γ be a nonempty subset of $\Omega = \{(3.3), (3.4), \ldots, (3.17)\}$, $\Gamma \neq \Omega$. Our objective here is to find t-norm T_1 and t-conorm C_1, dual operators, so that the equations in Γ are true using T_1 , C_1 and the equations not in Γ are false using T_1, C_1. Then find dual operators T_2 and C_2 so that the equations not in Γ are true using T_2 , C_2. We are to choose our t-norm and t-conorm from T_m, C_m, T_p, C_p, T_b, C_b, T^*, C^*. Then using T_1, C_1 on Γ and T_2, C_2 on $\Omega - \Gamma$ all the basic laws of crisp sets are true for fuzzy sets using mixed fuzzy logic.

For example, suppose we want to use T_m and C_m. We know that equations (3.10) and (3.11) are not true for T_m, C_m. We need to find T_2, C_2 from the set T_p,C_p, T_b, C_b, T^*, C^* so that using T_2, C_2 equations (3.10) and (3.11) are true. But also, if any equation is false for T_m, C_m it must be true for T_2, C_2.

There are only six possibilities: (1) $T_1 = T_m$, $C_1 = C_m$ and $T_2 = T_p$, $C_2 = C_p$; (2) $T_1 = T_m$, $C_1 = C_m$, and $T_2 = T_b$, $C_2 = C_b$; (3) $T_1 = T_m$, $C_1 = C_m$ and $T_2 = T^*$, $C_2 = C^*$; (4) $T_1 = T_p$, $C_2 = C_p$ and $T_2 = T_b$, $C_2 = C_b$; (5) $T_1 = T_p$, $C_1 = C_p$ and $T_2 = T^*$, $C_2 = C^*$; and (6) $T_1 = T_b$, $C_1 = C_b$ and $T_2 = T^*$, $C_2 = C^*$.

We claim at least one of these combinations will give us a mixed fuzzy logic so that all the basic formulas of crisp sets are also true for fuzzy sets. The exercises ask you to investigate all six cases.

As another application of mixed fuzzy logic suppose we wish to control the fuzziness of calculations like $\overline{A_1} \cap \overline{A_2} \cap \ldots \cap \overline{A_n}$ or $\overline{A_1} \cup \overline{A_2} \cup \ldots \cup \overline{A_n}$. The fuzziness of a fuzzy set may be measured by its support and the support of fuzzy set $\overline{B}$ is where $\overline{B}(x) > 0$. The support of $\overline{B}$, written $sp(\overline{B})$ is $\{x \in X | \overline{B}(x) > 0\}$.

Let $\overline{B} = \overline{A_1} \cap \ldots \cap \overline{A_n}$ and $\overline{C} = \overline{A_1} \cup \ldots \cup \overline{A_n}$. Now $sp(\overline{B})$ can vary depending on what t-norm we use to compute $\overline{B}$ and $sp(\overline{C})$ can change if we use different t-conorms to determine $\overline{C}$. We may think that the more fuzziness there is in $\overline{B}(\overline{C})$ the more uncertainty there is in the answer. So, suppose we want to choose t-norm T to minimize the fuzziness in $\overline{B}$, but we do not want $\overline{B} = \overline{\phi}$. Also, we wish to minimize the fuzziness in $\overline{C}$. We may use $T = T_m$, T_p, T_b, or T^* and $C = C_m$, C_p, T_b, or C^*.

Let $\overline{A_i}(x) = a_i$ in $[0,1]$, $1 \leq i \leq n$. Then $\overline{B}(x) = T(a_1, a_2, \ldots, a_n)$ and $\overline{C}(x) = C(a_1, \ldots, a_n)$. We will need to use the results in Section 3.3, and problems 7, 8 in that section. Let us look at an example using T_b and C_b. Then

$$\overline{B}(x) = T_b(a_1, \ldots, a_n) = \max(0, \sum_{i=1}^{n} a_i - n + 1). \tag{3.72}$$

The problem with T_b is that for larger and larger n, $\overline{B}$ tends towards $\overline{\phi}$. Suppose $\overline{A_i}(x) = 0.8$ all i and $n = 10$. Then

$$\overline{B}(x) = \max(0, -1) = 0. \tag{3.73}$$

Even when all the $\overline{A_i}$ have membership 0.8 at a given x, $\overline{B}(x) = 0$. The same is true for $\overline{A_i}(x) = 0.9$, $n = 10$. As you intersect more and more fuzzy sets using T_b you tend to get $\overline{\phi}$. We also see that

$$\overline{C}(x) = C_b(a_1, \ldots, a_n) = \min(1, \sum_{i=1}^{n} a_i). \tag{3.74}$$

Here we tend to get $\overline{C}(x) = 1$ all x, since the sum of the a_i may exceed one for large n. As you union more and more fuzzy sets using C_b you tend to get $\overline{X}$.

Table 3.1: Discrete Fuzzy Sets in Problem 3, Section 3.5.1

	0	1	2	3	4	5
$\overline{A_1}$	0	0.3	.7	1	0.6	0
$\overline{A_2}$	0	0	0.5	0.6	0.7	1
$\overline{A_3}$	1	0.5	0.3	0	0	0
$\overline{A_4}$	0.3	0.8	0	0	.8	.3

3.5.1 Exercises

1. We say a mixed fuzzy logic $((T_1, C_1), (T_2, C_2))$ is acceptable if: (1) for (T_1, C_1) the equations in Γ are true but those in $\Omega - \Gamma$ are false; and (2) for (T_2, C_2) those in $\Omega - \Gamma$ are true. Determine if the following are acceptable mixed fuzzy logics. In each case find Γ for (T_1, C_1):

   a. $((T_m, C_m), (T_p, C_p))$.

   b. $((T_m, C_m), (T_b, C_b))$.

   c. $((T_m, C_m), (T^*, C^*))$.

   d. $((T_p, C_p), (T_b, C_b))$.

   e. $((T_p, C_p), (T^*, C^*))$.

   f. $((T_b, C_b), (T^*, C^*))$.

2. Consider the three continuous fuzzy sets $\overline{A_1}$, $\overline{A_2}$, $\overline{A_3}$ in Figure 3.8.

   a. Find $\overline{B} = \overline{A_1} \cap \overline{A_2} \cap \overline{A_3}$ using $T = T_m, T_p, T_b$ and T^*.

   b. Is there any way to control for $sp(\overline{B})$? That is, do all values of T give the same value for $sp(\overline{B})$, or can we choose a unique T to minimize $sp(\overline{B})$?, $\overline{B} \neq \overline{\phi}$.

   c. Find $\overline{C} = \overline{A_1} \cup \overline{A_2} \cup \overline{A_3}$ using $C = C_m, C_p, C_b$ and C^*.

   d. Is there any way to control for $sp(\overline{C})$? Do all the C give the same $sp(\overline{C})$ or can you choose a unique C to minimize $sp(\overline{B})$?

   e. Repeat parts a through d using only $\overline{A_1}$ and $\overline{A_2}$.

3. In Table 3.1, $X = \{0, 1, 2, 3, 4, 5\}$, the values give the membership values for fuzzy sets $\overline{A_1}, \overline{A_2}, \overline{A_3}, \overline{A_4}$ at x in X.

   a. Find $\overline{B} = \overline{A_1} \cap \overline{A_2} \cap \overline{A_3} \cap \overline{A_4}$ using $T = T_m, T_p, T_b$ and T^*. Which T minimizes $sp(\overline{B})$?

   b. Find $\overline{C} = \overline{A_1} \cup \overline{A_2} \cup \overline{A_3} \cup \overline{A_4}$ using $C = C_m, C_p, C_b$ and C^*. Which C minimizes $sp(\overline{C})$?

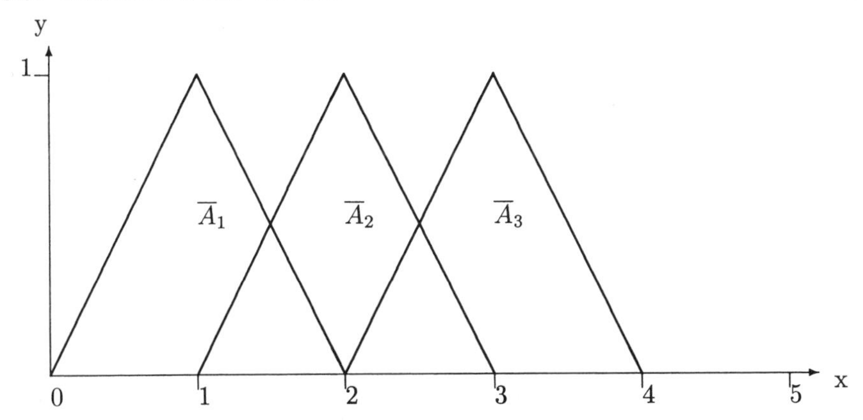

Figure 3.8: Continuous Fuzzy Sets in Problem 2, Section 3.5.1

4. Let $\overline{B} = \overline{A}_1 \cap \overline{A}_2$, the $\overline{A}_i$ continuous fuzzy subsets of the real numbers (like those in Figure 3.8). Give a general rule on which t-norm to use, $T = T_b, T_p, T_m, T^*$, to:

   a. minimize $sp(\overline{B})$, $\overline{B} \neq \overline{\phi}$.

   b. maximize $sp(\overline{B})$.

5. Let $\overline{C} = \overline{A}_1 \cup \overline{A}_2$, the $\overline{A}_i$ continuous fuzzy subsets of the real numbers (like those in Figure 3.8). Give a general rule on which t-conorm to use, C_b, C_p, C_m, C^*, to:

   a. minimize $sp(\overline{C})$.

   b. maximize $sp(\overline{C})$.

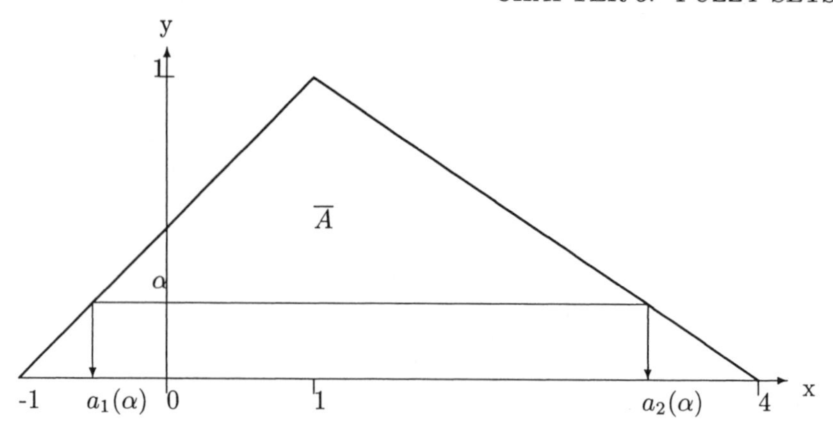

Figure 3.9: Continuous Fuzzy Set

3.6 Alpha-Cuts

If $\overline{A}$ is a fuzzy subset of universal set X, then the α-cut of $\overline{A}$, written $\overline{A}[\alpha]$, is defined as $\{x \in X|\overline{A}(x) \geq \alpha\}$, for $0 < \alpha \leq 1$. The $\alpha = 0$ cut, or $\overline{A}[0]$, must be defined separately because $\{x \in X|\overline{A}(x) \geq 0\}$ is always the whole universal set X. Notice that $\overline{A}[\alpha]$ is a crisp set for all α, $0 \leq \alpha \leq 1$.

The core of $\overline{A}$, written $co(\overline{A})$, is $\overline{A}[1]$ and the support of $\overline{A}$, $sp(\overline{A})$, is not $\overline{A}[0]$ but $\{x \in X|\overline{A}(x) > 0\}$. Notice that if $0 < \alpha_1 < \alpha_2 \leq 1$, then $\overline{A}[\alpha_2] \subseteq \overline{A}[\alpha_1]$. So the crisp sets $\overline{A}[\alpha]$ cannot increase as α increases from 0 to 1. The family of crisp sets $\overline{A}[\alpha]$, $0 < \alpha \leq 1$, are said to be a representation of the fuzzy set $\overline{A}$.

If $\overline{A}$ is a fuzzy subset of the real numbers, then we may separately define $\overline{A}[0]$ to be the closure of the support of $\overline{A}$. For example, if $sp(\overline{A})$ is the open interval $(1,5)$, then $\overline{A}[0] = [1,5]$ the closed interval. We may also take the closure of the support of $\overline{A}$ for $\overline{A}$ a fuzzy subset of $\mathbf{R} \times \mathbf{R}$. So, when we write $\overline{A}[\alpha]$, $0 \leq \alpha \leq 1$, we mean that $\overline{A}$ is a fuzzy subset of $\mathbf{R}$, or $\mathbf{R} \times \mathbf{R}$, $\cdots$ and $\overline{A}[0]$ is the closure of $sp(\overline{A})$. We will call $\overline{A}[0]$ the base of the fuzzy set $\overline{A}$.

Let us now find the α-cuts of some continuous and discrete fuzzy sets. Let $\overline{A}$ be the continuous fuzzy set in Figure 3.9. The α-cut, $0 \leq \alpha \leq 1$, gives an interval. From Figure 3.9, $\overline{A}[\alpha] = [a_1(\alpha), a_2(\alpha)]$ where $a_1(\alpha)$ is the left end point and $a_2(\alpha)$ is the right end point. The notation $x = a_i(\alpha)$, $0 \leq \alpha \leq 1$, $i = 1,2$ is backwards from the normal function notation. The usual functional notation is $y = f(x)$, or y is a function f of x. The notation $x = a_i(\alpha)$ is the inverse of $y = f(x)$ because α is in $[0,1]$ on the y-axis. In Figure 3.9 we see that $a_1(0) = -1, a_1(1) = 1$ and $a_2(1) = 1, a_2(0) = 4$. The properties of the $a_i(\alpha)$ functions are: $a_1(\alpha)$ is continuous and monotonically increasing from -1 to 1 on $[0,1]$; and (2) $a_2(\alpha)$ is continuous and monotonically decreasing from 4 to 1 on $[0,1]$.

We get the $a_i(\alpha)$ functions from first finding $y = f(x)$ for the left and right side of the membership function for $\overline{A}$. In Figure 3.9 we see that: (1) $y =$

$f_1(x) = \frac{1}{2}x + \frac{1}{2}, -1 \leq x \leq 1$, gives the left side; and (2) $y = f_2(x) = -\frac{1}{3}x + \frac{4}{3}$ for the right side. Now solve $\alpha = \frac{1}{2}x + \frac{1}{2}$ for x giving $x = 2\alpha - 1 = a_1(\alpha)$ and solve $\alpha = \frac{-1}{3}x + \frac{4}{3}$ for x producing $x = 4 - 3\alpha = a_2(\alpha)$. Hence $\overline{A}[\alpha] = [2\alpha - 1, 4 - 3\alpha], 0 \leq \alpha \leq 1$, for $\overline{A}$ in Figure 3.9. In general $a_1(\alpha) = f_1^{-1}(\alpha)$ and $a_2(\alpha) = f_2^{-1}(\alpha)$. If the core of $\overline{A}$ is not a single point, as in Figure 3.9, then we would require $a_1(1) < a_2(1)$ since $[a_1(1), a_2(1)] = co(\overline{A})$.

Find α-cuts of

$$\overline{A} = \{\frac{0}{x_1}, \frac{.7}{x_2}, \frac{1}{x_3}, \frac{.5}{x_4}, \frac{.2}{x_5}\}, \tag{3.75}$$

where $X = \{x_1, \ldots, x_5\}$. Then

$$\overline{A}[\alpha] = \{x_2, x_3, x_4, x_5\}, 0 < \alpha \leq 0.2, \tag{3.76}$$

$$\overline{A}[\alpha] = \{x_2, x_3, x_4\}, 0.2 < \alpha \leq 0.5, \tag{3.77}$$

$$\overline{A}[\alpha] = \{x_2, x_3\}, 0.5 < \alpha \leq 0.7, \tag{3.78}$$

$$\overline{A}[\alpha] = \{x_3\}, 0.7 < \alpha \leq 1. \tag{3.79}$$

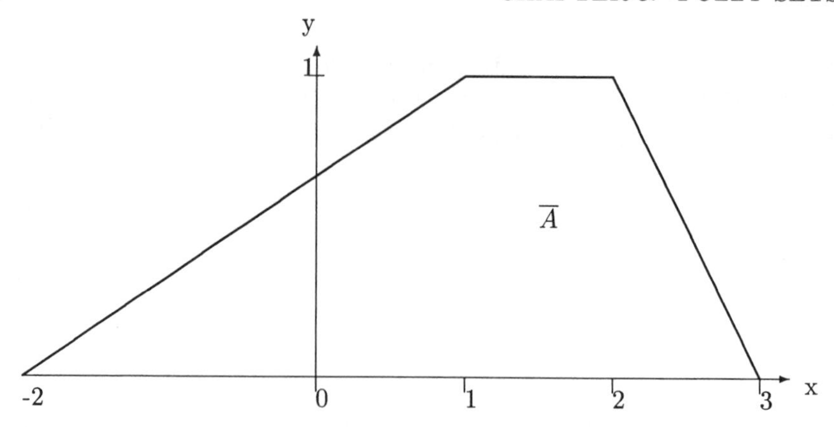

Figure 3.10: Continuous Fuzzy Set for Problem 8, Section 3.6.1

3.6.1 Exercises

1. Is $(\overline{A} \cap \overline{B})[\alpha] = \overline{A}[\alpha] \cap \overline{B}[\alpha]$ true or false when the t-norm T used to compute $\overline{A} \cap \overline{B}$ is

   a. $T = T_m$.

   b. $T = T_p$.

   c. $T = T_b$.

   d. $T = T^*$.

2. Is $(\overline{A} \cup \overline{B})[\alpha] = \overline{A}[\alpha] \cup \overline{B}[\alpha]$ true or false when the t-conorm C used to find $\overline{A} \cup \overline{B}$ is:

   a. $C = C_m$.

   b. $C = C_p$.

   c. $C = C_b$.

   d. $C = C^*$.

3. Is $(\overline{A}^c)[\alpha] = (\overline{A}[\alpha])^c$ true or false?

4. Show $\overline{A}[\alpha] = \bigcap\{\overline{A}[\beta] | \alpha < \beta \leq 1\}$, for $0 \leq \alpha < 1$.

5. If $\overline{A}$ is a crisp set, then what is $\overline{A}[\alpha], 0 < \alpha \leq 1$?

6. Show $\overline{A} \leq \overline{B}$ if and only if $\overline{A}[\alpha] \subseteq \overline{B}[\alpha]$, $0 < \alpha \leq 1$.

7. Show $\overline{A} = \overline{B}$ if and only if $\overline{A}[\alpha] = \overline{B}[\alpha], 0 < \alpha \leq 1$.

8. Find $\overline{A}[\alpha], 0 \leq \alpha \leq 1$, for $\overline{A}$ given in Figure 3.10.

9. Find $\overline{A}[\alpha], 0 \leq \alpha \leq 1$, for $\overline{A}$ given in Figure 3.11.

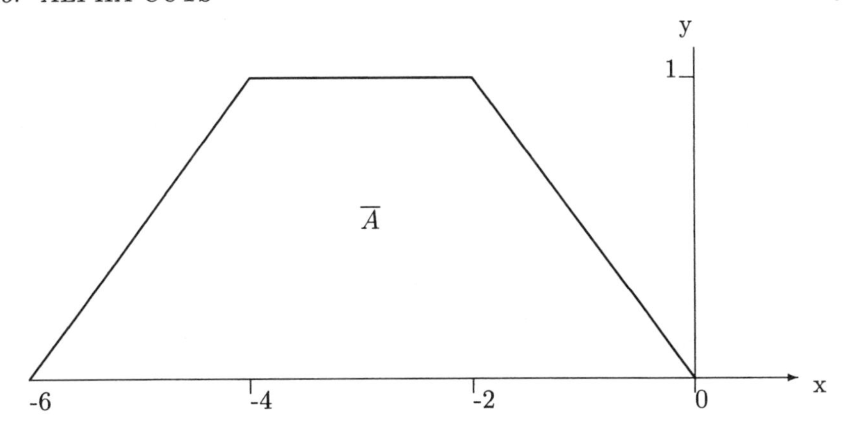

Figure 3.11: Continuous Fuzzy Set for Problem 9, Section 3.6.1

10. Give an example of a continuous fuzzy subset $\overline{A}$ of **R** where $\overline{A}[\alpha]$ is not an interval for all $0 \leq \alpha \leq 1$.

11. How would you define, if possible, α-cuts of type 2, or level 2, fuzzy sets?

12. How would you define α-cuts of fuzzy subsets of $\mathbf{R} \times \mathbf{R}$ (for example, fuzzy relations, Chapter 7)?

13. Find relationships, if any, between the sets $(\overline{A} \cap \overline{B})[\alpha]$ if we use t-norm T_m, T_b, T_p and T^* to calculate $\overline{A} \cap \overline{B}$.

14. Find relationships, if any, between the sets $(\overline{A} \cup \overline{B})[\alpha]$ if we use t-conorm C_m, C_b, C_p and C^* to find $\overline{A} \cup \overline{B}$.

15. Determine if the following equations are true or false. All fuzzy sets are fuzzy subsets of the real numbers:

    a.

$$\overline{A}^c[\alpha] \cup \overline{A}[\alpha] = \mathbf{R}.$$

    b.

$$\overline{A}^c[\alpha] \cap \overline{A}[\alpha] = \phi.$$

    c.

$$\overline{A}^c[\alpha] \cup \overline{B}^c[\alpha] = (\overline{A}[\alpha] \cap \overline{B}[\alpha])^c.$$

3.7 Distance Between Fuzzy Sets

Let $\mathcal{F}_0(\mathbf{R})$ be all continuous fuzzy subsets of $\mathbf{R}$ whose α-cuts are always bounded intervals. These will be called fuzzy numbers in the next chapter and are the fuzzy sets most used in applications. We need to be able to compute the distance between any $\overline{A}$ and $\overline{B}$ in $\mathcal{F}_0(\mathbf{R})$.

We know how to find the distance between two real numbers x, y. The distance is $|x - y| = d(x, y)$. We also know how to find the distance between two points in $\mathbf{R}^2$. The function d used to compute distance is called a metric. The basic properties of any metric for x, y in $\mathbf{R}$ are:

a. $d(x, y) \geq 0$;

b. $d(x, y) = d(y, x)$;

c. $d(x, y) = 0$ if and only if $x = y$; and

d. $d(x, y) \leq d(x, z) + d(z, y)$.

The first three properties of d are obvious: (1) distance is non-negative; (2) distance is symmetric; and (3) you get zero distance only when $x = y$. The fourth property says it is shorter to go directly to y from x instead of first going to intermediate point z.

Now we want a metric D for $\overline{A}, \overline{B}$ in $\mathcal{F}_0(\mathbf{R})$. D will have properties 1-4 given above. Also we point out that $D(\overline{A}, \overline{B})$ is a real number for $\overline{A}, \overline{B}$ fuzzy. Why do we want a metric for elements in $\mathcal{F}_0(\mathbf{R})$? We need it to start the calculus of fuzzy functions.

One usually starts calculus with the theory of limits. Consider a sequence x_n of real numbers. We say x_n converges to x if given $\epsilon > 0$ there is a N so that $d(x_n, x) < \epsilon, n \geq N$. To do this for a sequence $\overline{A_n}$, each $\overline{A_n}$ in $\mathcal{F}_0(\mathbf{R})$, we need a metric $D(\overline{A}, \overline{B}), \overline{A}, \overline{B}$ in $\mathcal{F}_0(\mathbf{R})$.

We now present two metrics for $\mathcal{F}_0(\mathbf{R})$. You are asked to investigate other possible metrics in the exercises.

Given $\overline{A}, \overline{B}$ in $\mathcal{F}_0(\mathbf{R})$, set $\overline{A}[\alpha] = [a_1(\alpha), a_2(\alpha)], \overline{B}[\alpha] = [b_1(\alpha), b_2(\alpha)], 0 \leq \alpha \leq 1$. Define $L(\alpha) = |a_1(\alpha) - b_1(\alpha)|, R(\alpha) = |a_2(\alpha) - b_2(\alpha)|$. Then

$$D(\overline{A}, \overline{B}) = \max\{\max(L(\alpha), R(\alpha)) | 0 \leq \alpha \leq 1\}. \tag{3.80}$$

Since $L(\alpha)$ and $R(\alpha)$ are continuous we used max instead of *sup*. This D is a metric.

Consider $\overline{A}$ and $\overline{B}$ in Figure 3.12. Then $a_1(\alpha) = 1 + \alpha$, $a_2(\alpha) = 4 - 2\alpha$, $b_1(\alpha) = 1 + 2\alpha$, $b_2(\alpha) = 4 - \alpha$. Then $L(\alpha) = \alpha$, $R(\alpha) = \alpha$, $\max(L(\alpha), R(\alpha)) = \alpha$ and $\max \alpha = 1$. This means $D(\overline{A}, \overline{B}) = 1$ for this metric.

Another metric is

$$D(\overline{A}, \overline{B}) = \int_a^b |\overline{A}(x) - \overline{B}(x)| dx, \tag{3.81}$$

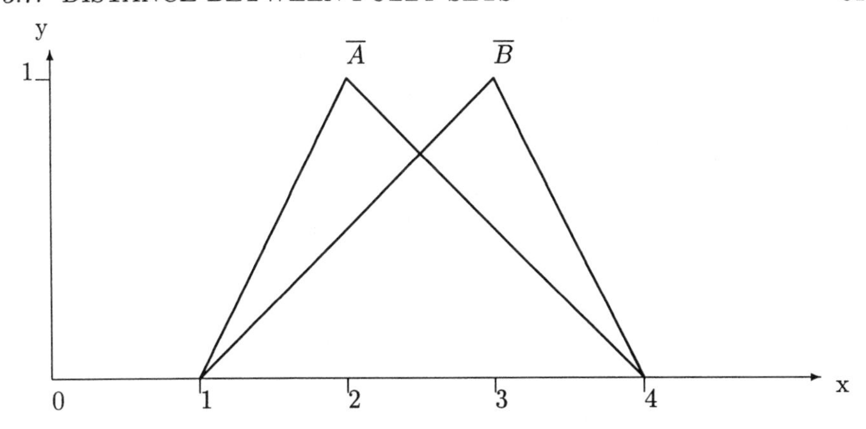

Figure 3.12: Two Continuous Fuzzy Numbers

where $[a,b]$ is an interval containing the support of $\overline{A}$ and $\overline{B}$. For the $\overline{A}$ and $\overline{B}$ in Figure 3.12 we first compute

$$|\overline{A}(x)-\overline{B}(x)| = \begin{cases} \frac{1}{2}x-\frac{1}{2} & \text{if } 1 \le x \le 2, \\ -x+\frac{5}{2} & \text{if } 2 \le x \le \frac{5}{2}, \\ x-\frac{5}{2} & \text{if } \frac{5}{2} \le x \le 3, \\ -\frac{1}{2}x+2 & \text{if } 3 \le x \le 4 \end{cases} \tag{3.82}$$

Then we integrate this giving $D(\overline{A},\overline{B})=\frac{3}{4}$.

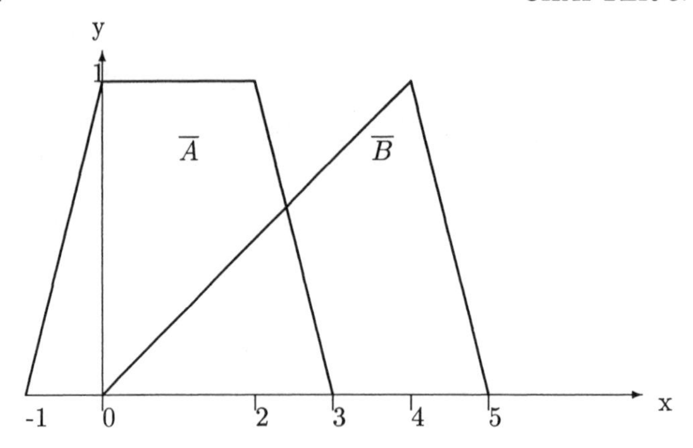

Figure 3.13: Fuzzy Sets in Problems 4 and 5

3.7.1 Exercises

1. For $\overline{A}, \overline{B}$ in $\mathcal{F}_0(\mathbf{R})$ define

$$D(\overline{A}, \overline{B}) = \max\{|A(x) - B(x)||x \in \mathbf{R}\}.$$

Is this D a metric? Justify your answer.

2. For $\overline{A}$ in $\mathcal{F}_0(\mathbf{R})$ define the center of $\overline{A}$, $cen(\overline{A})$, to be the center of the core of $\overline{A}$. Now set

$$D(\overline{A}, \overline{B}) = |cen(\overline{A}) - cen(\overline{B})|.$$

Is this D a metric? Justify your answer.

3. For $\overline{A}, \overline{B}$ in $\mathcal{F}_0(\mathbf{R})$ with $sp(\overline{A}) = (a_1, a_2), sp(\overline{B}) = (b_1, b_2)$, define

$$D(\overline{A}, \overline{B}) = |a_1 - b_1| + |a_2 - b_2|.$$

Is this D a metric? Justify your answer.

4. Given $\overline{A}$ and $\overline{B}$ in Figure 3.13, compute $D(\overline{A}, \overline{B})$ given in equation (3.80).

5. Given $\overline{A}$ and $\overline{B}$ in Figure 3.13, find $D(\overline{A}, \overline{B})$ in equation (3.81).

6. Explain why the D in equation (3.80) is called a horizontal metric while the D in equation (3.81) is called a vertical metric.

7. Convergence of sequences of continuous fuzzy numbers. This convergence is a simple extension of the definition of convergence of a sequence of real numbers given in the text. Determine if the sequence $\overline{A}_n$, $n = 1, 2, 3, \cdots$ converges and if it does converge, then find its limit $lim_{n\to\infty}\overline{A}_n = \overline{A}$. A triangular fuzzy number (section 4.2) is defined by its base and vertex. The notation $(a/b/c)$ is for a triangular fuzzy number (like $\overline{A}$ and $\overline{B}$ in Figure 3.12) with base the interval $[a, c]$ and vertex at $x = b$.

    a. $\overline{A}_n = (\frac{2-n}{2+n}/1/1 + (0.1)^n)$
    b. $\overline{A}_n = (c_n - \frac{1}{n}/c_n/c_n + \frac{1}{n})$ for $c_n = (1 + \frac{r}{n})^n$, r a real number.
    c. $\overline{A}_n = ([ln(n)/n]/1/n)$.
    d. $\overline{A}_n = (\frac{1}{3}^n/n^{1/n}/tan^{-1}(n))$.
    e. $\overline{A}$ is a triangular fuzzy number, $\overline{A}_n = \overline{A} + r\overline{A} + r^2\overline{A} + \cdots + r^{n-1}\overline{A}$ for $r \in (0, 1)$.

8. Define $D(\overline{A}, \overline{B}) = 0$ if $\overline{A} = \overline{B}$ and the value is 1 otherwise, for $\overline{A}$ and $\overline{B}$ in $\mathcal{F}_0(\mathbf{R})$. Is this D a metric?

9. Define
$$D(\overline{A}, \overline{B}) = sup\{|\overline{A}(x) - \overline{B}(x)||x \in \mathbf{R}\}.$$
Is this D a metric for $\overline{A}$ and $\overline{B}$ in $\mathcal{F}_0(\mathbf{R})$?

10. Find $D(\overline{A}, \overline{A}^c)$ for $\overline{A}$ given in Figure 3.13 using:

    a. Equation (3.80);
    b. Equation (3.81).

Chapter 4

Fuzzy Numbers

4.1 Introduction

Fuzzy numbers are of great importance in fuzzy systems. In the next section we give the general definition of a fuzzy number and show that real numbers and closed intervals are special cases of fuzzy numbers. The fuzzy numbers usually used in applications are the triangular (shaped) and the trapezoidal (shaped) fuzzy numbers also defined in section two. Then, in the third section, we develop the arithmetic of fuzzy numbers using both the extension principle and the α-cut with interval arithmetic method. There are many properties of the real numbers we use frequently that can be translated over to fuzzy numbers. These include : (1) finding the distance between two real numbers, becoming the distance between fuzzy numbers discussed in Section 3.7; (2) finding the maximum and the minimum of two real numbers, resulting in fuzzy max and fuzzy min in Section 4; and (3) ordering real numbers, or writing $x \leq y$ and $x > y$, extended to $\overline{M} \leq \overline{N}$ and $\overline{M} > \overline{N}$ for fuzzy numbers $\overline{M}$ and $\overline{N}$ in Section 5. Certain fuzzy systems (Chapter 14) produce a fuzzy set as its output and we can not use this fuzzy set to communicate to a machine. When this happens the fuzzy set is defuzzified, or mapped to a real number, and the resulting real number (for speed, voltage, etc.) is then sent to the machine. Defuzzification is discussed in the last section.

4.2 Fuzzy Numbers

Fuzzy numbers are very special fuzzy subsets of the real numbers. For example, a fuzzy number expressing approximately 2 is in Figure 4.1 and another fuzzy number (also called a fuzzy interval) for approximately between 2 and 4 is shown in Figure 4.2.

The general definition of a fuzzy number $\overline{N}$ is a fuzzy subset of $\mathbf{R}$ and:

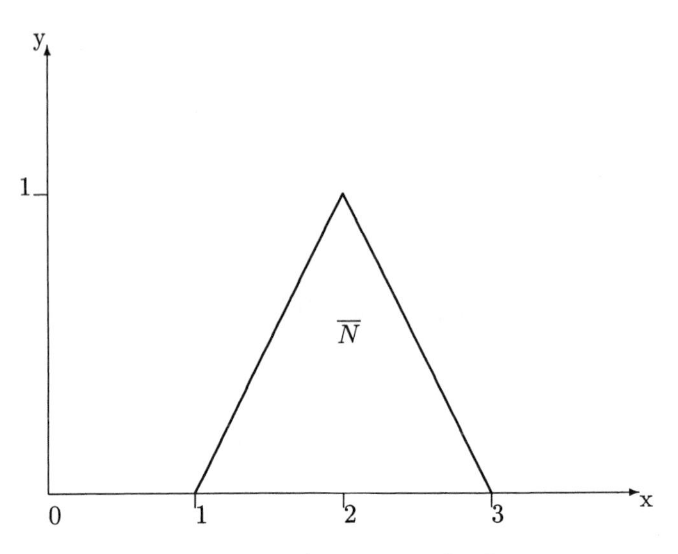

Figure 4.1: Approximately Two

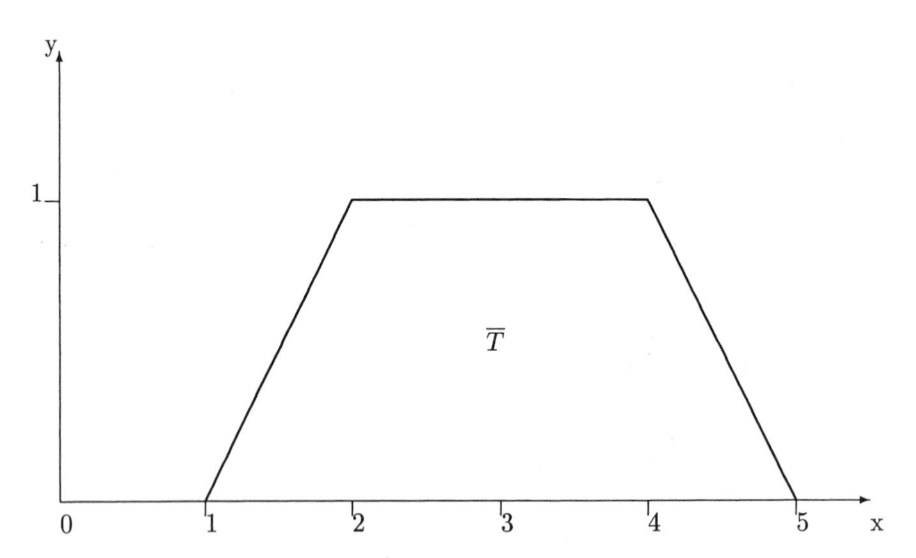

Figure 4.2: Approximately Two to Four

1. the core of $\overline{N}$ is non-empty;
2. α-cuts of $\overline{N}$ are all closed, bounded, intervals; and
3. the support of $\overline{N}$ is bounded.

The core of $\overline{N}$ is $\overline{N}[1]$, the $\alpha = 1$ cut. So $\overline{N}$ must be a normal fuzzy set. The α-cuts of $\overline{N}$, $\overline{N}[\alpha]$, are always $[n_1(\alpha),\, n_2(\alpha)]$, $0 \leq \alpha \leq 1$. $\overline{N}[\alpha]$ is a closed, bounded, interval for all $0 \leq \alpha \leq 1$ means that $\overline{N}[\alpha] = [n_1(\alpha), n_2(\alpha)]$ where $n_1(\alpha)$ is the left end point and $n_2(\alpha)$ is the right end point of this interval. The support of $\overline{N}$, $\{x|\overline{N}(x) > 0\}$, bounded means that there is a positive number M so that the support of $\overline{N}$ is a subset of $[-M, M]$. We do not allow intervals like (a, ∞), $(-\infty, b)$ or $(-\infty, \infty)$ for the support of $\overline{N}$. We purposely did not say that the membership function had to be continuous.

Consider the fuzzy set $\overline{N}$ in Figure 4.3. The support of $\overline{N}$ is the interval $(1, 6)$ and the core of $\overline{N}$ is the interval $[3, 4]$. Notice the jump discontinuity at $x = 2$. The point $(2, 0.4)$ is not on the graph but the point $(2, 0.6)$ is on the graph. It must be this way for $\overline{N}[\alpha]$ to always be a closed interval.

Continuing our discussion of the fuzzy number in Figure 4.3 let us first find x a function of α for x in the intervals $[1, 2)$, $[2, 3]$ and $[4, 6]$. We did this in section 3.6 of Chapter 3. You first find y a function of x, substitute α for y and then solve for x. We obtain: (1) $x = (5/2)\alpha + 1$, $0 \leq \alpha < 0.4$, $1 \leq x < 2$; (2) $x = (5/2)\alpha + (1/2)$, $0.6 \leq \alpha \leq 1$, $2 \leq x \leq 3$; and (3) $x = 6 - 2\alpha$, $0 \leq \alpha \leq 1$, $4 \leq x \leq 6$. Now we may find $\overline{N}[\alpha]$, $0 \leq \alpha \leq 1$. We see that:

1. $\overline{N}[0] = [1, 6]$;
2. $\overline{N}[0.2] = [1.5, 5.6]$;
3. $\overline{N}[0.4] = [2, 5.2]$;
4. $\overline{N}[0.5] = [2, 5]$;
5. $\overline{N}[0.6] = [2, 4.8]$;
6. $\overline{N}[0.8] = [2.5, 4.4]$; and
7. $\overline{N}[1] = [3, 4]$.

They are all closed intervals. If we moved the point on the graph at $x = 2$ from $(2, 0.6)$ to $(2, 0.4)$ we would get $\overline{N}[0.4] = [2, 5.2]$, $\overline{N}[0.5] = (2, 5]$ and $\overline{N}[0.6] = (2, 4.8]$ not all closed intervals.

Why do we allow such $\overline{N}$ as in Figure 4.3 to be called a fuzzy number? So that all real numbers and all crisp closed intervals become fuzzy numbers. Let $\mathcal{N}$ be the set of all fuzzy numbers, let $x = r$ be a real number, and let $I = [a, b]$ be a closed interval. A fuzzy number $\overline{M}$ equal to r is

$$\overline{M}(x) = \begin{cases} 1, & \text{if } \; x = r \\ 0, & \text{otherwise.} \end{cases} \tag{4.1}$$

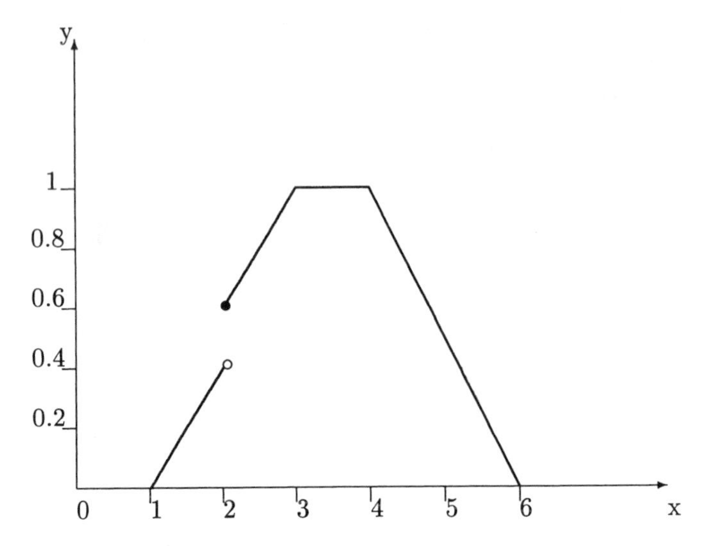

Figure 4.3: A General Fuzzy Number

$\overline{M}$ is a fuzzy number because its core is $[r, r]$, its support is $[r, r]$ and all its α-cuts are $[r, r]$ all closed intervals. So we say r belongs to $\mathcal{N}$.

Let $\overline{P}$ be the fuzzy number

$$\overline{P}(x) = \begin{cases} 1, & \text{if} \quad x \in [a, b] \\ 0, & \text{otherwise,} \end{cases} \tag{4.2}$$

so that $\overline{P}$ equals I. Hence, I also belongs to $\mathcal{N}$. Therefore, $\mathbf{R}$ is a subset of $\mathcal{N}$ and the set of all closed intervals is also a subset of $\mathcal{N}$.

The general definition of a fuzzy number was to generalize crisp numbers and crisp closed intervals. However, in applications we hardly ever use an $\overline{N}$ as shown in Figure 4.3. In applications we use continuous fuzzy numbers. We now define the two basic types of fuzzy numbers used in practice.

A triangular fuzzy number $\overline{N}$, Figure 4.1, is defined by three numbers $a < b < c$ where the vertex of the triangle is at $x = b$ and its base is the interval $[a, c]$. We write $\overline{N} = (a/b/c)$ for a triangular fuzzy number. A triangular shaped fuzzy number $\overline{M}$ is partially defined by $a < b < c$ because its sides are continuous curves, not straight lines. We write $\overline{M} \approx (1/2/3)$ for a triangular shaped fuzzy number $\overline{M}$ shown in Figure 4.4. A trapezoidal fuzzy number $\overline{T}$ is defined by four number $a < b < c < d$ where $\overline{T}(x) = 1$ on $[b, c]$ and its base is the interval $[a, d]$ as shown in Figure 4.2. We write $\overline{T} = (a/b, c/d)$ for a trapezoidal fuzzy number. A trapezoidal shaped fuzzy number $\overline{U}$ is partially defined by $a < b < c < d$, because its sides need not be straight lines as shown in Figure 4.5. We write $\overline{U} \approx (1/2, 3/4)$ for a trapezoidal shaped fuzzy number shown in Figure 4.5. We will usually be using triangular (shaped) and trapezoidal (shaped) fuzzy numbers in this book. For $\overline{N}$ and $\overline{M}$ define $\overline{N}[0] = \overline{M}[0] = [a, c]$ and also define $\overline{T}[0] = \overline{U}[0] = [a, d]$, called the base of

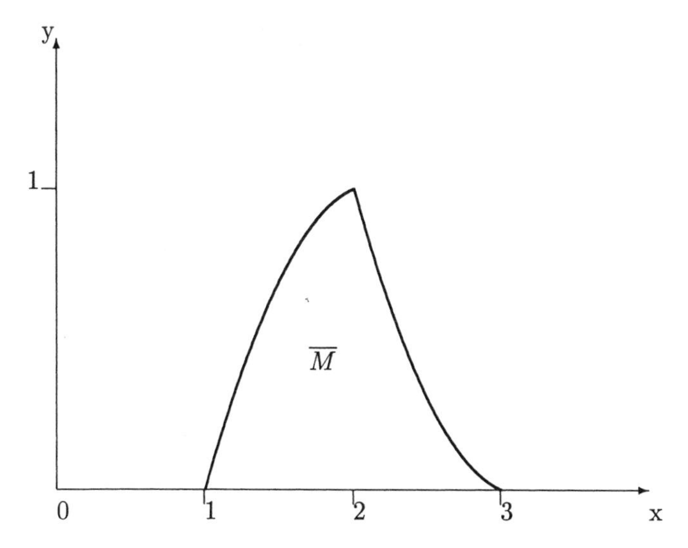

Figure 4.4: Triangular Shaped Fuzzy Number

the fuzzy number. For fuzzy numbers the $\alpha = 0$ cut is the base of the fuzzy number.

If $\overline{N} = (a/b/c)$, or $\overline{N} \approx (a/b/c)$, we write: (1) $\overline{N} > 0$ if $a > 0$; (2) $\overline{N} \geq 0$ if $a \geq 0$; (3) $\overline{N} < 0$ when $c < 0$; and (4) $\overline{N} \leq 0$ whenever $c \leq 0$. Similarly we define $\overline{T} > 0$, $\overline{T} \geq 0$, $\overline{T} < 0$ and $\overline{T} \leq 0$ for $\overline{T} = (a/b, c/d)$ or $\overline{T} \approx (a/b, c/d)$. It is now obvious how we would define $\overline{N} > r$, $\overline{T} \leq r$, etc. for real number r.

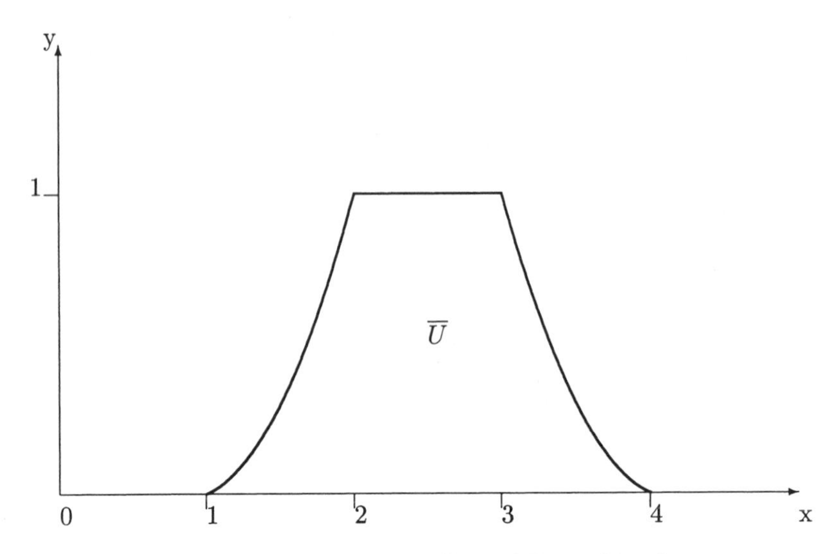

Figure 4.5: Trapezoidal Shaped Fuzzy Number

4.2.1 Exercises

1. Let $\overline{N}(x) = \exp(x)$, $x \leq 0$ and $\overline{N}(x) = \exp(-x)$ for $x \geq 0$. Is $\overline{N}$ a fuzzy number?

2. Let $\overline{N}(x) = 1/x, x \geq 1$ and $\overline{N}(x) = (2-x)^{-1}$, $x \leq 1$. Is $\overline{N}$ a fuzzy number?

3. Let $\overline{N}(x) = \sin(x), 0 \leq x \leq \pi$ and $\overline{N}(x) = 0$ otherwise. Is $\overline{N}$ a fuzzy number?

4. Let $\overline{N}(x) = x$, for $0 \leq x \leq 1$ and 0 otherwise. Is $\overline{N}$ a fuzzy number?

5. Let $\overline{N}(x) = 1$ for $x = 0, 1, 2, 3, 4, 5$ and $\overline{N}(x) = 0$ otherwise. Is $\overline{N}$ a fuzzy number?

6. Let $I = (3, 7)$ the open interval from 3 to 7. Can you define a fuzzy number $\overline{M}$ so that $\overline{M}$ equals I?

7. Let $I = (-\infty, 3]$. Can you define a fuzzy number $\overline{V}$ so that $\overline{V}$ equals I?

8. For the fuzzy number $\overline{W}$ in Figure 4.6 find $\overline{W}[\alpha]$ for $\alpha = 0.3, 0.6, 0.9$.

9. For the fuzzy number $\overline{Y}$ in Figure 4.7 find $\overline{Y}[\alpha]$ for $\alpha = 0.3, 0.4, 0.5, 0.7$.

10. For any fuzzy number $\overline{N}$ find its height $ht(\overline{N})$.

11. If $\overline{N}$ is a fuzzy number, is $\overline{N}^c$ a fuzzy number?

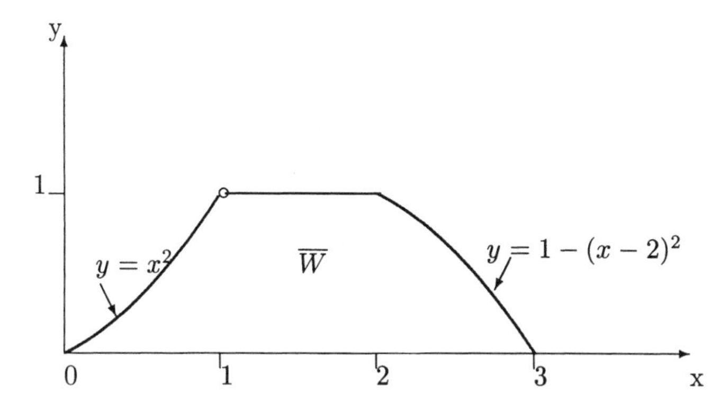

Figure 4.6: Fuzzy Number for Problem 8

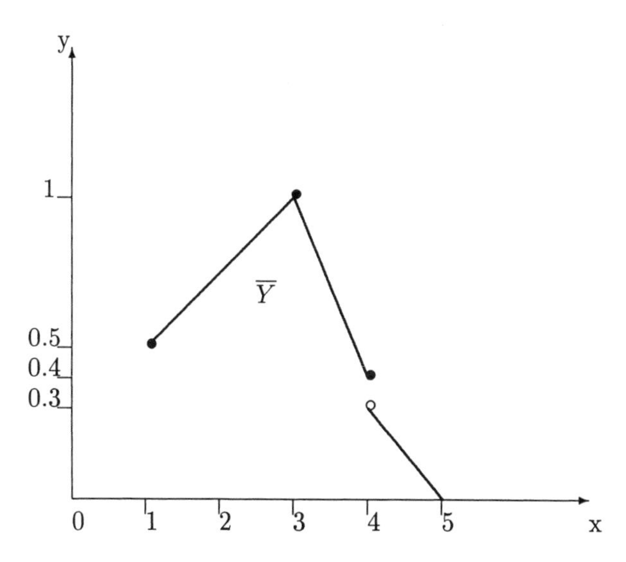

Figure 4.7: Fuzzy Number for Problem 9

12. If $\overline{N}$ and $\overline{M}$ are two fuzzy numbers is $\overline{N} \cup \overline{M}$ and/or $\overline{N} \cap \overline{M}$ a fuzzy number when we use t-norm T_m for intersection and t-conorm C_m for union? Will either one be a fuzzy number if we switch to T_b and C_b, or T_p,C_p, or T^*,C^*?

13. Let $\overline{D}$ be a discrete fuzzy subset of **R**. When can $\overline{D}$ be a fuzzy number?

14. Draw a picture of two continuous fuzzy numbers $\overline{N}$ and $\overline{M}$, both approximately equal to 10, so that $\overline{N} \leq \overline{M}$.

15. Draw a picture of two continuous fuzzy numbers $\overline{N}$ and $\overline{M}$, both approximately between -6 to -4 so that $\overline{N} \leq \overline{M}$.

16. Let $\overline{A}$ and $\overline{B}$ be the triangular fuzzy numbers shown in Figure 3.4.

    a. Is $\overline{C}_1$ in Figure 3.5 a fuzzy number?

    b. Is $\overline{D}_1$ in Figure 3.6 a fuzzy number?

    c. Is $\overline{D}_2$ in Figure 3.7 a fuzzy number?

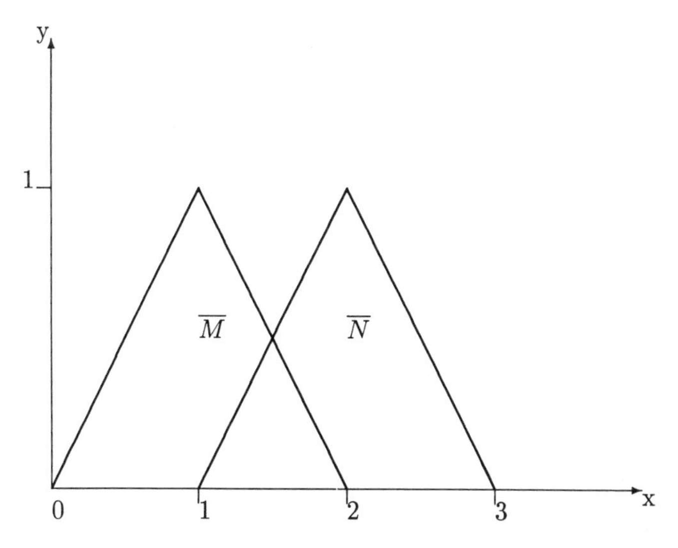

Figure 4.8: Finding $\overline{P} = \overline{M} + \overline{N}$

4.3 Fuzzy Arithmetic

Let $\mathcal{N}$ be all fuzzy numbers. Given $\overline{N}$ and $\overline{M}$ in $\mathcal{N}$ we wish to compute $\overline{N} + \overline{M}$, $\overline{N} - \overline{M}$, $\overline{N} \cdot \overline{M}$ and $\overline{N} \div \overline{M}$. There are two basic ways to do this: (1) using the extension principle; and (2) using α-cuts and interval arithmetic. We will discuss the extension principle first. The extension principle will be used more extensively in Chapters 5 and 8.

4.3.1 Extension Principle

Let $\overline{M}$ and $\overline{N}$ be in $\mathcal{N}$ and let $\overline{P} = \overline{M} + \overline{N}$. The extension principle method of finding the membership function for $\overline{P}$ is

$$\overline{P}(z) = \sup\{\min(\overline{M}(x), \overline{N}(y)) | x + y = z\}. \tag{4.3}$$

Let us illustrate the computations involved in equation (4.3) using $\overline{M}$ and $\overline{N}$ in Figure 4.8. We first pick a value for z, say $z = 2$, then evaluate $\min(\overline{M}(x), \overline{N}(y))$ for x and y which add up to $z = 2$. We have done this for certain values of x and y as shown in Table 4.1. It looks like the max occurs for $x = 0.5$ and $y = 1.5$ so that $\overline{P}(2) = 0.5$. Now do this for other values for z. $\overline{P}(3)$ is easy to find since $x + y = 3$ for $x = 1, y = 2$ and $\overline{M}(1) = \overline{N}(2) = 1$ and then $\overline{P}(3) = 1$. When we are all finished we get $\overline{P} = (1/3/5)$ a triangular fuzzy number.

For subtraction let $\overline{Q} = \overline{M} - \overline{N}$, then

$$\overline{Q}(z) = \sup\{\min(\overline{M}(x), \overline{N}(y)) | x - y = z\}. \tag{4.4}$$

Table 4.1: Finding the Sum of Two Fuzzy Numbers

x	$\overline{M}(x)$	y	$\overline{N}(y)$	min(col#2,col#4)
1	1	1	0	0
0.9	0.9	1.1	0.1	0.1
0.7	0.7	1.3	0.3	0.3
0.5	0.5	1.5	0.5	0.5
0.3	0.3	1.7	0.7	0.3
0.1	0.1	1.9	0.9	0.1
0	0	2	1	0

Table 4.2: Finding the Product of Two Fuzzy Numbers

x	$\overline{M}(x)$	y	$\overline{N}(y)$	min(col#2,col#4)
2	0	1	0	0
1.75	0.25	8/7	1/7	1/7
1.5	0.50	4/3	1/3	1/3
1.25	0.75	8/5	3/5	3/5
1	1	2	1	1
0.75	0.75	8/3	1/3	1/3
0.50	0.50	4	0	0
0.25	0.25	8	0	0

If $\overline{R} = \overline{M} \cdot \overline{N}$, then its membership function is

$$\overline{R}(z) = \sup\{\min(\overline{M}(x), \overline{N}(y)) | xy = z\}. \tag{4.5}$$

We know that if $\overline{M}$ and $\overline{N}$ are triangular, or trapezoidal, fuzzy numbers, then so is $\overline{M} + \overline{N}$ and $\overline{M} - \overline{N}$. However, $\overline{M} \cdot \overline{N}$ will be a triangular, or trapezoidal, shaped fuzzy number if both are triangular, or trapezoidal, fuzzy numbers. Let us find $\overline{R}(2)$ for $\overline{M}$ and $\overline{N}$ in Figure 4.8. Table 4.2 gives some computational results. It looks like $\overline{R}(2) = 1$. In fact $\overline{R} \approx (0/2/6)$, a triangular shaped fuzzy number.

Lastly, let $\overline{S} = \overline{M}/\overline{N}$. We now have to assume that zero does not belong to the interval $\overline{N}[0]$. Then

$$\overline{S}(z) = \sup\{\min(\overline{M}(x), \overline{N}(y)) | x/y = z\}. \tag{4.6}$$

Find $\overline{M}/\overline{N}$ for $\overline{M}$ and $\overline{N}$ given in Figure 4.8 (see the Exercises at the end of this section).

Notice that in equations (4.3)–(4.6) we always used t-norm $T = T_m = \min$. Other t-norms can be used. Let $\star$ denote $\pm$, $\cdot$, or $\div$. Then if $\overline{P} = \overline{M} \star \overline{N}$, we can find $\overline{P}$ as follows

$$\overline{P}(z) = \sup\{T(\overline{M}(x), \overline{N}(y)) | x \star y = z\}, \tag{4.7}$$

for t-norm T.

To control the fuzziness in the result (see Section 3.5), we might choose $T \neq T_m$. The fuzziness in a $\overline{P}$ in $\mathcal{N}$ could be measured by the length of the interval $\overline{P}[0]$ and/or the size of $\overline{P}[1]$. For example, using $T = T_m$ and computing $\overline{P} = \overline{M} + \overline{N}$ for $\overline{M}$ and $\overline{N}$ in Figure 4.8, the fuzziness ($\overline{P}[0]$) equals 4 while the fuzziness (base) of $\overline{M}$ and $\overline{N}$ is only 2. To reduce the fuzziness, or uncertainty, in the result try using another t-norm.

Exercises

1. Using $T = T_m$ find $\overline{M} - \overline{N}$ for $\overline{M}$ and $\overline{N}$ given in Figure 4.8. Let $\overline{Q} = \overline{M} - \overline{N}$. Find $\overline{Q}(z)$ at $z = -3, -2, -1, 0, 1$. Draw a picture of $\overline{Q}$. Find $a < b < c$ so that $\overline{Q} = (a/b/c)$ or $\overline{Q} \approx (a/b/c)$.

2. Using $T = T_m$ find $\overline{M}/\overline{N}$ for $\overline{M}$ and $\overline{N}$ in Figure 4.8. Let $\overline{S} = \overline{M}/\overline{N}$. Find $\overline{S}(z)$ at $z = 0, 0.25, 0.50, 1, 1.5, 2$. Draw a picture of $\overline{S}$. Find $a < b < c$ so that $\overline{S} = (a/b/c)$ or $\overline{S} \approx (a/b/c)$.

3. Using $T = T^*$ let $\overline{P} = \overline{M} + \overline{N}$, with $\overline{M}$ and $\overline{N}$ in Figure 4.8. Find $\overline{P}(z)$ for $z = 1, 2, 3, 4, 5$. Draw a picture of $\overline{P}$. Can you find $a < b < c$ so that $\overline{P} = (a/b/c)$ or $\overline{P} \approx (a/b/c)$?.

4. Use $\overline{N}$ from Figure 4.8 and use t-norm T_m.

   a. Find $3 + \overline{N}$. Determine $a < b < c$ so that $3 + \overline{N} = (a/b/c)$.

   b. Find $2\overline{N}$. Does $2\overline{N} = \overline{N} + \overline{N}$? Can you find $a < b < c$ so that $2\overline{N} = (a/b/c)$ or $2\overline{N} \approx (a/b/c)$?

   c. Find $\overline{N} \div 2$ and $2 \div \overline{N}$. Are they triangular (shaped) fuzzy numbers?

5. Redo Problem 4 using $T = T_p$.

6. Let $\overline{M} = (-1/0, 1/2)$ and $\overline{N} = (4/6/7)$. Compute the lengths of the intervals $(\overline{M} \star \overline{N})[0]$ and $(\overline{M} \star \overline{N})[1]$ for $\star = \pm, \cdot, \div$ and for $T = T_m, T_b, T_p, T^*$. Which t-norm minimizes the length of these intervals?

7. Let $\overline{P}, \overline{M}, \overline{N}$ be fuzzy numbers. Determine if the following equation is true or false (use t-norm T_m for intersection and t-conorm C_m for union):

   a. $\overline{P} + (\overline{M} \cap \overline{N}) = (\overline{P} + \overline{M}) \cap (\overline{P} + \overline{N})$.

   b. $\overline{P} + (\overline{M} \cup \overline{N}) = (\overline{P} + \overline{M}) \cup (\overline{P} + \overline{N})$.

   In addition, now also assume that $\overline{M} \leq \overline{N}$.

   c. $\overline{P} + \overline{M} \leq \overline{P} + \overline{N}$.

   d. $\overline{P} - \overline{M} \leq \overline{P} - \overline{N}$.

   e. $\overline{P} \cdot \overline{M} \leq \overline{P} \cdot \overline{N}$.

   f. $\overline{P} \div \overline{M} \leq \overline{P} \div \overline{N}$.

8. Use $\overline{N}$ from Figure 4.8 and use t-norm T_m.

   a. Find $[-4, -1] + \overline{N}$. Determine $a < b < c$ so that $[-4, -1] + \overline{N} = (a/b/c)$.

b. Find $[-4,-1]\overline{N}$. Can you find $a < b < c$ so that $[-4,-1]\overline{N} = (a/b/c)$ or $[-4,-1]\overline{N} \approx (a/b/c)$?

c. Find $\overline{N} \div [-4,-1]$ and $[-4,-1] \div \overline{N}$. Are they triangular (shaped) fuzzy numbers?

9. Redo Problem 8 using $T = T^*$.

10. If $\overline{M}$ and $\overline{N}$ are arbitrary fuzzy numbers, is $\overline{M} \star \overline{N}$, for $\star = \pm, \cdot, \div$, always a fuzzy number?

11. Show that zero always belongs to the support of $\overline{M} - \overline{M}$ and that one always belongs to the support of $\overline{M} \div \overline{M}$.

12. Let $X = \{x_1, x_2, x_3\}$ and

$$\overline{A} = \{\frac{0.6}{x_1}, \frac{1}{x_2}, \frac{0.2}{x_3}\},$$

$$\overline{B} = \{\frac{0.8}{x_1}, \frac{0.4}{x_2}, \frac{1}{x_3}\}.$$

Give numerical values to the x_i like $x_1 = -3$, $x_2 = 1$ and $x_3 = 4$. Find, using t-norm T_m, $\overline{A} \pm \overline{B}$, $\overline{A}\,\overline{B}$ and $\overline{A} \div \overline{B}$.

13. Redo Problem 1 using T_b.

14. Redo Problem 1 using T^*.

15. Redo Problem 2 using T_b.

16. Redo Problem 2 using T^*.

17. Redo Problem 3 using T_b.

18. Redo Problem 3 using T_p.

4.3.2 Interval Arithmetic

Before we can proceed to the second method of doing fuzzy arithmetic we first have to learn some of the basics of interval arithmetic. Let $I = [a, b]$ and $J = [c, d]$ be two closed intervals. If $\star = \pm, \cdot, \div$ (we also use "/" for $\div$), then

$$I \star J = \{x \star y | x \in I, y \in J\}. \tag{4.8}$$

For example, if $I = [1, 3]$ and $J = [2, 5]$ and $\star$ is subtraction, then

$$[1, 3] - [2, 5] = \{x - y | x \in [1, 3], y \in [2, 5]\}, \tag{4.9}$$

which equals $[-4, 1]$. You get the end points from $1 - 5 = -4$ and $3 - 2 = 1$.

The results from equation (4.8) can be summarized as follows:

1. $[a, b] + [c, d] = [a + c, b + d]$;
2. $[a, b] - [c, d] = [a - d, b - c]$;
3. $[a, b] \cdot [c, d] = [\alpha, \beta]$;
   $\alpha = \min\{ac, ad, bc, bd\}$;
   $\beta = \max\{ac, ad, bc, bd\}$;
4. $[a, b] \div [c, d] = [a, b] \cdot [1/d, 1/c]$,

as long as zero does not belong to $[c, d]$ when we divide by this interval.

Multiplication can be simplified if you know that the intervals are positive or negative. For example, if $a \geq 0$ and $c \geq 0$, then

$$I \cdot J = [ac, bd], \tag{4.10}$$

and if $b < 0$ and $c \geq 0$, then

$$I \cdot J = [ad, cb]. \tag{4.11}$$

An interval $[r, r]$ can be identified with the real number r. Let $1^* = [1, 1]$ and $0^* = [0, 0]$. So we have $I + 0^* = I$ and $I \cdot 1^* = I$, etc. and we have the algebra of intervals. Certain properties of real numbers also hold for intervals but some properties of real numbers do not hold for intervals.

Exercises

1. Let $I = [a, b]$ and $J = [c, d]$. Find $I \cdot J$ if:

   a. $b < 0$ and $d < 0$;

   b. $a \geq 0$ and $d < 0$;

   c. $a \geq 0$, $c < 0$ and $d \geq 0$;

   d. $a < 0$, $b \geq 0$ and $c \geq 0$;

   e. $a < 0$, $b \geq 0$, $c < 0$ and $d \geq 0$;

   f. $a < 0$, $b \geq 0$ and $d < 0$;

   g. $b < 0$, $c < 0$ and $d \geq 0$.

2. Let $I = [a, b]$ and $J = [c, d]$. Find $I \div J$ if:

   a. $a \geq 0$ and $c > 0$;

   b. $b < 0$ and $c > 0$;

   c. $a \geq 0$, $c < 0$ and $d \geq 0$;

   d. $a < 0$, $b \geq 0$ and $c \geq 0$;

   e. $a < 0$, $b \geq 0$, $c < 0$ and $d \geq 0$;

   f. $a < 0$, $b \geq 0$ and $d < 0$;

   g. $b < 0$, $c < 0$ and $d \geq 0$.

3. Let I, J and K be closed intervals. If the following equation is true, then give an argument showing that it is true. If it is false, then give an example which shows it is not true.

   a. $I + (J + K) = (I + J) + K$.

   b. There is a closed interval L so that $I + L = 0$, where 0 is the interval $[0, 0] = 0$.

   c. $I \cdot (J \cdot K) = (I \cdot J) \cdot K$.

   d. There is a closed interval Y so that $I \cdot Y = 1$, where 1 is the interval $[1, 1] = 1$.

   e. $I \cdot (J + K) = I \cdot J + I \cdot K$.

   f. If $I \subseteq X$ and $J \subseteq Y$, where both X and Y are closed intervals, then $I \cdot J \subseteq X \cdot Y$ and $I/J \subseteq X/Y$.

4. Show that the real number zero always belongs to $I - I$.

5. Show that the real number one always belongs to I/I.

6. Given an equation $I + X = J$ can you always solve for closed interval X?

7. Given the equation $I \cdot X = J$ can you always solve for closed interval X?

8. Consider the closed interval $I = [a, b]$ and $J = [c, d]$ as fuzzy subsets of $\mathbf{R}$. Use the extension principle to find $I \star J$ for $\star \in \{\pm, \cdot, \div\}$. Are the results the same as in interval arithmetic? Explain.

4.3.3 Alfa-Cuts and Interval Arithmetic

Alfa-cuts (α-cuts) of fuzzy numbers are always closed and bounded intervals. If $\overline{M}$ and $\overline{N}$ are two fuzzy numbers let $\overline{M}[\alpha] = [m_1(\alpha), m_2(\alpha)]$ and $\overline{N}[\alpha] = [n_1(\alpha), n_2(\alpha)]$, for $0 \leq \alpha \leq 1$. Then we may define the arithmetic of fuzzy numbers in terms of their α-cuts. If $\overline{P} = \overline{M} + \overline{N}$, then

$$\overline{P}[\alpha] = \overline{M}[\alpha] + \overline{N}[\alpha], \tag{4.12}$$

or

$$\overline{P}[\alpha] = [m_1(\alpha) + n_1(\alpha), m_2(\alpha) + n_2(\alpha)], \tag{4.13}$$

for $0 \leq \alpha \leq 1$.

We can do the same for subtraction. Let $\overline{Q} = \overline{M} - \overline{N}$ and then we see that

$$\overline{Q}[\alpha] = [m_1(\alpha) - n_2(\alpha), m_2(\alpha) - n_1(\alpha)], \tag{4.14}$$

for all α. If $\overline{R} = \overline{M} \cdot \overline{N}$, then

$$\overline{R}[\alpha] = [m_1(\alpha), m_2(\alpha)] \cdot [n_1(\alpha), n_2(\alpha)], \tag{4.15}$$

for $0 \leq \alpha \leq 1$. Lastly, if $\overline{S} = \overline{M}/\overline{N}$, then

$$\overline{S}[\alpha] = [m_1(\alpha), m_2(\alpha)] \cdot [1/n_2(\alpha), 1/n_1(\alpha)], \tag{4.16}$$

assuming that zero does not belong to $\overline{N}[0]$.

Equations (4.12)–(4.16) give the α-cuts of $\overline{P}, \overline{Q}, \overline{R}, \overline{S}$. You put these α-cuts together to get the fuzzy sets $\overline{P}, \overline{Q}, \overline{R}, \overline{S}$. As an example of the computation let us find these four fuzzy numbers for the $\overline{M}$ and $\overline{N}$ in Figure 4.8. First you need to find $\overline{M}[\alpha] = [\alpha, 2 - \alpha]$ and $\overline{N}[\alpha] = [1 + \alpha, 3 - \alpha]$. Then

1. $\overline{P}[\alpha] = [1 + 2\alpha, 5 - 2\alpha]$,
2. $\overline{Q}[\alpha] = [-3 + 2\alpha, 1 - 2\alpha]$,
3. $\overline{R}[\alpha] = [\alpha + \alpha^2, 6 - 5\alpha + \alpha^2]$, and
4. $\overline{S}[\alpha] = [\alpha/(3 - \alpha), (2 - \alpha)/(1 + \alpha)]$.

We see that $\overline{P} = (1/3/5)$, $\overline{Q} = (-3/-1/1)$, $\overline{R} \approx (0/2/6)$ and $\overline{S} \approx (0/0.5/2)$. In order to sketch their graphs you need to find the inverse functions.

Let us first graph $\overline{Q}$. We start with $-3 + 2\alpha = x$ and $1 - 2\alpha = x$. Solve for α giving $y = \alpha = (x + 3)/2$ and $y = \alpha = (1 - x)/2$. The graph of $\overline{Q}$ is in Figure 4.9.

Let us next do the same for $\overline{R}$. Solve $\alpha^2 + \alpha = x$ for α giving $y = \alpha = (-1 + \sqrt{1 + 4x})/2 = f_1(x)$ and solve $6 - 5\alpha + \alpha^2 = x$ for α producing $y = \alpha = (5 - \sqrt{1 + 4x})/2 = f_2(x)$. In each case we had a choice of $\pm\sqrt{1 + 4x}$. In the first case we picked $+$ so that $x = 0$ gave $y = 0$. In the second case

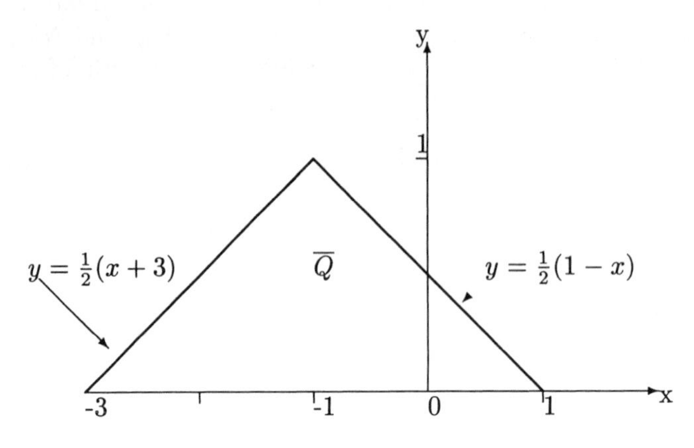

Figure 4.9: $\overline{Q} = \overline{M} - \overline{N}$

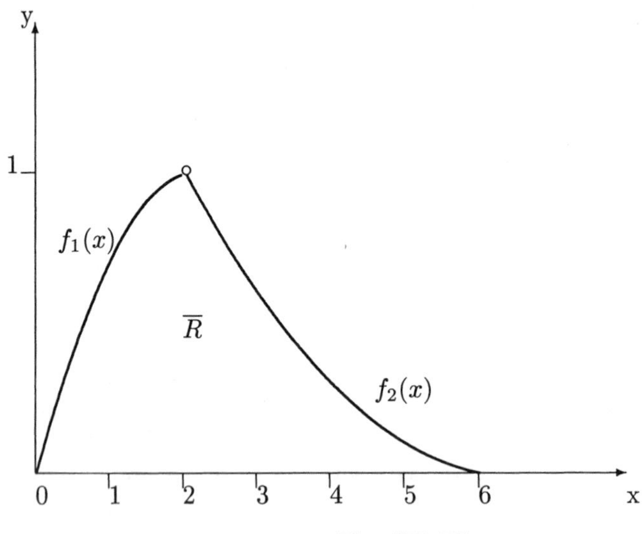

Figure 4.10: $\overline{R} = \overline{M} \cdot \overline{N}$

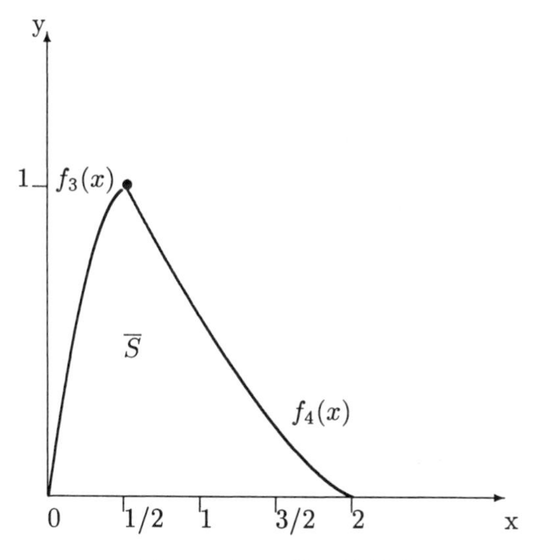

Figure 4.11: $\overline{S} = \overline{M}/\overline{N}$

choose - (minus) since $x = 6$ is to produce $y = 0$. The graph of $\overline{R}$ is in Figure 4.10.

Let us also graph $\overline{S}$. Solve $\alpha/(3-\alpha) = x$ for α giving $y = \alpha = f_3(x) = 3x/(1+x)$ and solve $(2-\alpha)/(1+\alpha) = x$ for α producing $y = \alpha = f_4(x) = (2-x)/(1+x)$. $\overline{S}$ is in Figure 4.11.

Clearly, this method of fuzzy arithmetic seems much easier than the extension principle procedure of Section 4.3.1. This method is also easily implemented on a computer. We would discretize the fuzzy numbers by only computing their α-cuts for certain values of α, say for $\alpha = 0, 0.1, \ldots, 0.9, 1$. Then, using equations (4.12)–(4.16) we could quickly, using interval arithmetic, find $\overline{P}, \overline{Q}, \overline{R}, \overline{S}$. But, do the two methods of doing fuzzy arithmetic give the same results? It is well-known that for continuous fuzzy numbers $\overline{M}$ and $\overline{N}$ the two procedures, using t-norm $T_m = \min$ in equation (4.7) of Section 4.3.1, and the α-cut and interval arithmetic method of this section, give the same results for $\overline{P}, \overline{Q}, \overline{R}, \overline{S}$.

Exercises

1. Let $\overline{M} = (-3/-1/1)$ and $\overline{N} = (1/3/4)$. Find α-cuts of the following fuzzy sets and sketch their graphs. In all cases find $y = f(x)$ for the left and right sides of the fuzzy sets as in Figures 4.9–4.11.

   a. $\overline{P} = \overline{M} + \overline{N}$.
   b. $\overline{Q} = \overline{M} - \overline{N}$.
   c. $\overline{R} = \overline{M} \cdot \overline{N}$.
   d. $\overline{S} = \overline{M}/\overline{N}$.

2. Let $\overline{M} = (-4/-3, -1/0)$ and $\overline{N} = (3/4, 5/6)$. Same instructions as for Problem 1.

3. Let $\overline{M} = (a/b/c)$ and r is a real number.

   a. Does $r + \overline{M} = (a + r/b + r/c + r)$?
   b. If $r > 0$, does $r \cdot \overline{M} = (ar/br/cr)$?
   c. If $r < 0$, does $r \cdot \overline{M} = (cr/br/ar)$?

4. Let $\overline{M}$ be a fuzzy number.

   a. If $\overline{Q} = \overline{M} - \overline{M}$, then find $\overline{Q}[0]$.
   b. If $\overline{S} = \overline{M}/\overline{M}$, then find $\overline{S}[1]$.

5. If $\overline{M} \geq 0$, then how would you define $\overline{M}^r$ for $r = 2, 3, 4, \ldots$?

6. How would you define $\exp(\overline{M})$.

7. Given $\overline{M} > 0$, how would you define $\ln(\overline{M})$?

8. Can you give a definition for $\sin(\overline{M})$?

9. Can you give a definition of $\tan(\overline{M})$?

10. Compute $\pm, \cdot, \div$ for discrete fuzzy sets

$$\overline{M} = \{\frac{0.2}{1}, \frac{0.4}{2}, \frac{0}{3}, \frac{0.3}{4}, \frac{0.7}{5}\},$$

and

$$\overline{N} = \{\frac{0}{1}, \frac{0.4}{2}, \frac{0.7}{3}, \frac{0}{4}, \frac{0.8}{5}\}.$$

11. Do the two methods, the extension principle and the α-cut with interval arithmetic, produce the same results for all fuzzy numbers?

12. Rework the following problems in the Exercises in Section 4.3.1 using the α-cut and interval arithmetic method of this section.

a. Problem 1.

b. Problem 2.

c. Problem 4.

d. Problem 8.

e. Problem 12.

13. Suppose in section 4.3.1 and equations (4.3)–(4.6) we use another t-norm T. Do the two methods, extension principle and α-cuts with interval arithmetic, still give the same results for $\pm, \cdot, \div$ for continuous fuzzy numbers? Investigate for:

    a. $T = T_b$,

    b. $T = T_p$, and

    c. $T = T^*$.

4.3.4 Properties of Fuzzy Arithmetic

In this section let us be working with only continuous fuzzy numbers. They are all the triangular (shaped) and trapezoidal (shapes) fuzzy numbers. Let $\overline{M}, \overline{N}, \overline{P}$ be three continuous fuzzy numbers.

The basic properties we wish to investigate are those of the real numbers translated into continuous fuzzy numbers. The first set of basic properties involve addition and subtraction.

1. (closure) $\overline{M} + \overline{N}$ is a continuous fuzzy number.
2. (commutativity) $\overline{M} + \overline{N} = \overline{N} + \overline{M}$.
3. (associativity) $\overline{M} + (\overline{N} + \overline{P}) = (\overline{M} + \overline{N}) + \overline{P}$.
4. (additive identity) There is a continuous fuzzy set $\overline{O}$ so that $\overline{M} + \overline{O} = \overline{M}$.
5. (additive inverse) There is a continuous fuzzy set $\overline{X}$ so that $\overline{M} + \overline{X} = \overline{O}$.

The question is: "which of these are true for all $\overline{M}, \overline{N}, \overline{P}$?" We can use either method, the extension principle or the α-cut with interval arithmetic, to show they are true or false. Let us look at numbers 2 and 3 using the α-cut and interval arithmetic method. Clearly number 2 is true because $I + J = J + I$ is true for intervals I and J. Number 3 is true or false depending on how you answered Problem 3a in the Exercises to section 4.3.2. Notice if we allow all fuzzy numbers so that the real number 0 (zero) is a fuzzy number, then number 4 is true because $\overline{M} + 0 = \overline{M}$.

The next set of basic properties involves multiplication and division:

1. (closure) $\overline{M} \cdot \overline{N}$ is a continuous fuzzy number,
2. (commutativity) $\overline{M} \cdot \overline{N} = \overline{N} \cdot \overline{M}$,
3. (associativity) $\overline{M} \cdot (\overline{N} \cdot \overline{P}) = (\overline{M} \cdot \overline{N}) \cdot \overline{P}$,
4. (multiplicative identity) There is a continuous fuzzy number $\overline{1}$ so that $\overline{M} \cdot \overline{1} = \overline{M}$, and
5. (multiplicative inverse) There is a continuous fuzzy number $\overline{Y}$ so that $\overline{M} \cdot \overline{Y} = \overline{1}$.

Using α-cuts and interval arithmetic we know that number 2 is true because $I \cdot J = J \cdot I$ is true for intervals. Property 3 for intervals was answered in problem 3c in the Exercises to Section 4.3.2. If we allow all fuzzy numbers then the real number 1 (one) is a fuzzy number and property 4 is true since $\overline{M} \cdot 1 = \overline{M}$.

The last basic property involves the interaction of addition and multiplication and it is called the distributive law

$$\overline{M} \cdot (\overline{N} + \overline{P}) = \overline{M} \cdot \overline{N} + \overline{M} \cdot \overline{P}. \tag{4.17}$$

The distributive law was looked at for intervals in problem 3e in the Exercises to Section 4.3.2.

It is known that closure (addition and multiplication) is true for continuous fuzzy numbers. You are now asked if the rest of the basic properties are true for continuous fuzzy numbers.

Exercises

1. Determine whether or not the following property is true for continuous fuzzy numbers:

   a. $\overline{M} \cdot (\overline{N} \cdot \overline{P}) = (\overline{M} \cdot \overline{N}) \cdot \overline{P}$,

   b. $\overline{M} + (\overline{N} + \overline{P}) = (\overline{M} + \overline{N}) + \overline{P}$,

   c. There is an $\overline{O}$ so that $\overline{M} + \overline{O} = \overline{M}$ for all $\overline{M}$,

   d. For each $\overline{M}$ there is an $\overline{X}$ so that $\overline{M} + \overline{X} = \overline{O}$,

   e. There is a $\overline{1}$ so that $\overline{M} \cdot \overline{1} = \overline{M}$ for all $\overline{M}$, and

   f. For each $\overline{M}$ there is a $\overline{Y}$ so that $\overline{M} \cdot \overline{Y} = \overline{1}$.

2. Show that

$$\overline{M} \cdot (\overline{N} + \overline{P}) \leq (\overline{M} \cdot \overline{N}) + (\overline{M} \cdot \overline{P})$$

   is true for all continuous fuzzy numbers.

3. Determine if the distributive law is true if

   a. $\overline{M}$ is the real number r (so $\overline{M}$ is not a continuous fuzzy number),

   b. $\overline{N} > 0$ and $\overline{P} > 0$,

   c. $\overline{N} < 0$ and $\overline{P} < 0$, and

   d. Any $\overline{M}$, $\overline{N}$, and $\overline{P}$.

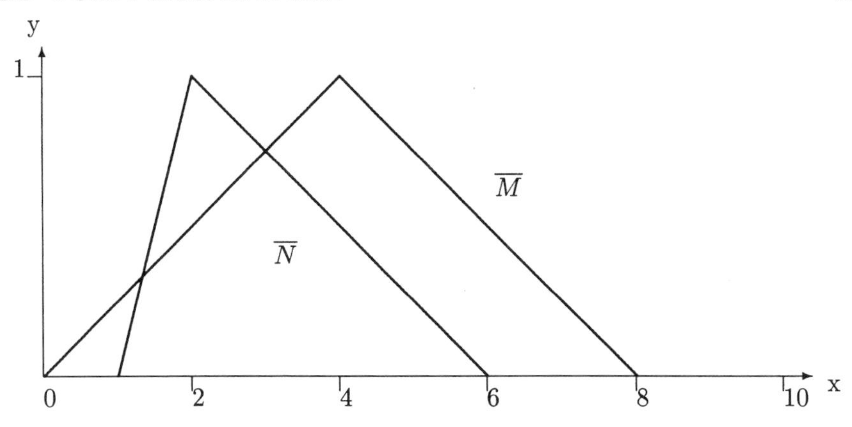

Figure 4.12: Fuzzy Numbers $\overline{M}$ and $\overline{N}$

4.4 Fuzzy Max and Min

We are familiar with $\max\{x, y\}$ and $\min\{x, y\}$ for real numbers x, y. We now fuzzify max to fuzzy max, written $\overline{\max}$, and we fuzzify min to fuzzy min, written $\overline{\min}$. Let $\overline{M}$ and $\overline{N}$ be two continuous fuzzy numbers. If $\overline{P} = \overline{\max}(\overline{M}, \overline{N})$, then we find the membership function for $\overline{P}$ using the extension principle as follows

$$\overline{P}(z) = \sup\{\min(\overline{M}(x), \overline{N}(y)) | \max\{x, y\} = z\}, \tag{4.18}$$

Also, if we let $\overline{Q} = \overline{\min}(\overline{M}, \overline{N})$ then

$$\overline{Q}(z) = \sup\{\min(\overline{M}(x), \overline{N}(y)) | \min\{x, y\} = z\}. \tag{4.19}$$

Example 4.4.1

Let us find $\overline{P}$ and $\overline{Q}$ for the two triangular fuzzy numbers $\overline{M} = (0/4/8)$ and $\overline{N} = (1/2/6)$ shown in Figure 4.12. Tables 4.3 and 4.4 show some of the calculations for $\overline{P}(2)$ and $\overline{Q}(2)$. Tables 4.5 and 4.6 have computations for $\overline{P}(4)$ and $\overline{Q}(4)$. From Table 4.3 it looks like $\overline{P}(2) = 0.5$ and Table 4.4 implies that $\overline{Q}(2) = 1$. See Figures 4.13 and 4.14. Table 4.5 implies that $\overline{P}(4) = 1$ and it looks like $\overline{Q}(4) = 0.5$ from Table 4.6.

We notice that fuzzy max ($\overline{\max}$) and fuzzy min ($\overline{\min}$) are quite different from crisp (non-fuzzy) max and min. The difference is when $\overline{M}$ and $\overline{N}$ overlap, or when $\overline{M} \cap \overline{N} \neq \overline{\phi}$ using t-norm min, $\overline{\max}(\overline{M}, \overline{N})$ and $\overline{\min}(\overline{M}, \overline{N})$ do not equal $\overline{M}$ or $\overline{N}$. However, if $\overline{M} \cap \overline{N} = \overline{\phi}$ then fuzzy max (min) can equal $\overline{M}$ or $\overline{N}$. This is discussed in the exercises.

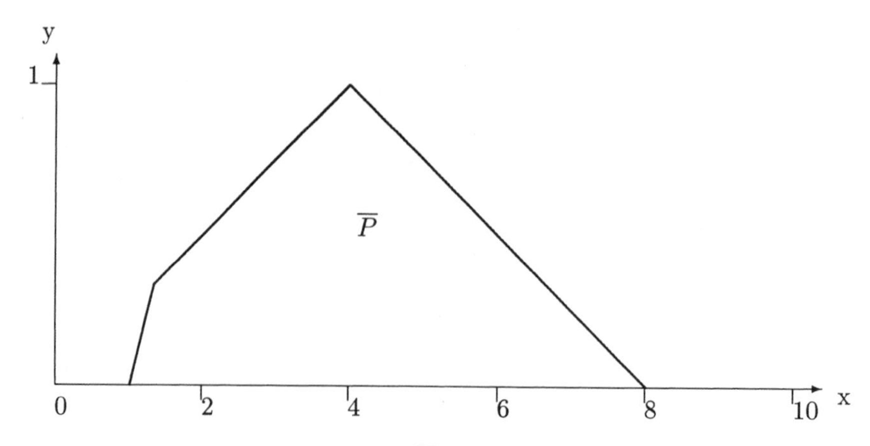

Figure 4.13: $\overline{P} = \overline{max}(\overline{M}, \overline{N})$

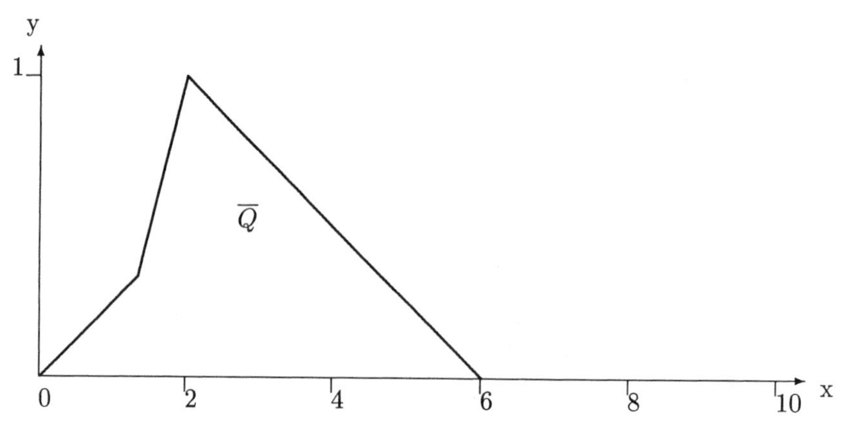

Figure 4.14: $\overline{Q} = \overline{\min}(\overline{M}, \overline{N})$

Table 4.3: Finding Fuzzy Max at $z = \max(x, y) = 2$

x	y	$\overline{M}(x)$	$\overline{N}(y)$	min(col #3, col #4)
2	0	0.5	0	0
2	1	0.5	0	0
2	2	0.5	1	0.5
0	2	0	1	0
1	2	0.25	1	0.25

Table 4.4: Finding Fuzzy Min at $z = \min(x, y) = 2$

x	y	$\overline{M}(x)$	$\overline{N}(y)$	min(col#3,col#4)
2	6	0.5	0	0
2	5	0.5	0.25	0.25
2	4	0.5	0.5	0.5
2	3	0.5	0.75	0.5
2	2	0.33	1	0.33
3	2	0.75	1	0.75
4	2	1	1	1
5	2	0.75	1	0.75
6	2	0.5	1	0.5
7	2	0.25	1	0.25
8	2	0	1	0

Table 4.5: Finding Fuzzy Max at $z = \max(x, y) = 4$

x	y	$\overline{M}(x)$	$\overline{N}(y)$	min(col#3,#4)
4	4	1	0.5	0.5
4	3	1	0.75	0.75
4	2	1	1	1
4	1	1	0	0
3	4	0.75	0.5	0.5
2	4	0.5	0.5	0.5
1	4	0.25	0.5	0.25
0	4	0	0.5	0

Table 4.6: Finding Fuzzy Min at $z = \min(x, y) = 4$

x	y	$\overline{M}(x)$	$\overline{N}(y)$	min(col#3,col#4)
4	4	1	0.5	0.5
4	5	1	0.25	0.25
4	6	1	0	0
3	4	0.75	0.5	0.5
2	4	0.5	0.5	0.5
1	4	0.25	0.5	0.25
0	4	0	0.5	0

4.4.1 Exercises

1. Suppose $\overline{M} \cap \overline{N} = \overline{\phi}$ using t-norm min. Show that $\overline{\max}(\overline{M}, \overline{N})$ equals $\overline{M}$ or $\overline{N}$ and if fuzzy max is $\overline{M}$ ($\overline{N}$), the fuzzy min equals $\overline{N}$ ($\overline{M}$).

2. Suppose $\overline{M} \leq \overline{N}$, meaning $\overline{M}(x) \leq \overline{N}(x)$ all x. Draw a picture of the fuzzy max and the fuzzy min of $\overline{M}$ and $\overline{N}$.

3. Show that $\overline{\max}(\overline{M}, \overline{M}) = \overline{M}$ and $\overline{\min}(\overline{M}, \overline{M}) = \overline{M}$.

4. Show that $\overline{\max}(\overline{M}, \overline{N}) = \overline{\max}(\overline{N}, \overline{M})$ and $\overline{\min}(\overline{M}, \overline{N}) = \overline{\min}(\overline{N}, \overline{M})$.

5. If $\overline{\max}(\overline{M}, \overline{N}) = \overline{M}$, then is it true that

$$\overline{\max}(\overline{M} + \overline{P}, \overline{N} + \overline{P}) = \overline{M} + \overline{P}$$

for any continuous fuzzy number $\overline{P}$?

6. Let $\overline{P} = \overline{\max}(\overline{M}, \overline{N})$ and $\overline{P}[\alpha] = [p_1(\alpha), p_2(\alpha)]$ all α. Can we find these α-cuts as follows

$$p_1(\alpha) = \min\{\max(a, b) | a \in \overline{M}[\alpha], b \in \overline{N}[\alpha]\}$$

and

$$p_2(\alpha) = \max\{\max(a, b) | a \in \overline{M}[\alpha], b \in \overline{N}[\alpha]\}.$$

7. In the definition of fuzzy max let us use another t-norm T

$$\overline{P}(z) = \sup\{T(\overline{M}(x), \overline{N}(y)) | \max(x, y) = z\},$$

so that $\overline{P} = \overline{\max}(\overline{M}, \overline{N})$ using t-norm T. Rework Example 4.4.1 using:

   a. $T = T_b$,
   b. $T = T_p$, and
   c. $T = T^*$.

8. Repeat Problem 7 for fuzzy min.

9. Let $\overline{M} = (1/2/3)$ and $\overline{N} = (2/3/4)$. Use t-norm min and draw pictures of

   a. $\overline{\max}(\overline{M}, \overline{N})$,
   b. $\overline{\min}(\overline{M}, \overline{N})$.

10. Repeat Problem 9 for $\overline{M} = (-1/0/2)$ and $\overline{N} = (1/3/4)$.

11. Let $X = \{0, 1, 2, 3, 4, 5\}$ and

$$\overline{A} = \{\frac{0.4}{0}, \frac{0}{1}, \frac{0.8}{2}, \frac{1}{3}, \frac{0.5}{4}, \frac{0.2}{5}\},$$

$$\overline{B} = \{\frac{0.7}{0}, \frac{0.5}{1}, \frac{1}{2}, \frac{0.6}{3}, \frac{0.1}{4}, \frac{0.8}{5}\}.$$

Find the fuzzy max and the fuzzy min of $\overline{A}$ and $\overline{B}$.

4.5 Inequalities

In this section we wish to define "<" and "≤" between fuzzy numbers and discuss their properties.

Up to now "≤" between fuzzy numbers meant "fuzzy subset". $\overline{M} \leq \overline{N}$ meant that $\overline{M}$ was a fuzzy subset of $\overline{N}$, or $\overline{M}(x) \leq \overline{N}(x)$, for all x. In this section ≤ will mean "less than or equal to". In the rest of the book when "≤" appears it should be clear, from how it is used, which meaning is attached to the symbol; if not we will state the meaning.

The symbol "≤ " is made up of two parts "<" and "=". We will use "<" between two fuzzy numbers but we will not use "=" between fuzzy numbers because $\overline{M} = \overline{N}$ implies that $\overline{M}$ and $\overline{N}$ are identically equal. Instead of "=" we will use "≈" to mean that the two fuzzy numbers are approximately equal or identical. So, $\overline{M} \approx \overline{N}$ means $\overline{M} = \overline{N}$, or $\overline{M}$ and $\overline{N}$ are approximately equal. Then "≤" will now mean "<" or "≈".

A total (linear, complete) order "≤" on the set of fuzzy numbers has the following four properties:

1. (reflexive) $\overline{M} \leq \overline{M}$;
2. (transitive) $\overline{M} \leq \overline{N}$ and $\overline{N} \leq \overline{P}$, implies $\overline{M} \leq \overline{P}$;
3. $\overline{M} \leq \overline{N}$ and $\overline{N} \leq \overline{M}$ implies $\overline{M} \approx \overline{N}$;
4. For any two fuzzy numbers $\overline{M}, \overline{N}$ we have $\overline{M} \leq \overline{N}$ or $\overline{N} \leq \overline{M}$.

If "≤" only has the first three properties, then it is called a partial order on the set of fuzzy numbers.

The usual ordering on the real numbers is a total ordering. So, we would like to have a total ordering on the set of fuzzy numbers. If we can not get an "acceptable" total ordering, then we will settle for a partial ordering.

A common method of defining ≤ on the set of fuzzy numbers (Example 4.5.1 below) is first to define the meaning of $\overline{M} < \overline{N}$. In this method $\overline{M} < \overline{M}$ is always false, and $\overline{M} < \overline{N}$ and $\overline{N} < \overline{M}$ can not both be true. Then $\overline{M} \approx \overline{N}$ whenever $\overline{M} < \overline{N}$ is not true and $\overline{N} < \overline{M}$ is false. So $\overline{M} \leq \overline{N}$ means $\overline{M} < \overline{N}$ or $\overline{M} \approx \overline{N}$. Notice that when we specify ≤ this way properties #1, #3 and #4 are all automatically satisfied. All we need to do is check to see if transitivity is also satisfied for ≤ to be a total ordering. But that is the problem because transitivity usually fails to hold. Transitivity usually fails because: (1) < is transitive ($\overline{M} < \overline{N}$ and $\overline{N} < \overline{P}$ implies $\overline{M} < \overline{P}$); but (2) ≈ is not transitive ($\overline{M} \approx \overline{N}$ and $\overline{N} \approx \overline{P}$ but $\overline{M}$ is not approximately equal to $\overline{P}$). Now ≈ must be transitive for ≤ to be transitive (see the exercises).

The relation ≈ fails to be transitive because we may have $\overline{M} \approx \overline{N}$ and $\overline{N} \approx \overline{P}$ but $\overline{M} < \overline{P}$. We can get this result because: (1) $\overline{N}$ lies a little to the right of $\overline{M}$ so that $\overline{M} \approx \overline{N}$; (2) $\overline{P}$ lies a little to the right of $\overline{N}$ so that $\overline{N} \approx \overline{P}$; but (3) $\overline{P}$ lies far enough to the right of $\overline{M}$ so that $\overline{M} < \overline{P}$.

Let us now look at some definitions of ≤ on the set of fuzzy numbers.

Example 4.5.1

We first define $<$ between fuzzy numbers $\overline{M}$ and $\overline{N}$. Let

$$v(\overline{M} \leq \overline{N}) = \sup\{\min(\overline{M}(x), \overline{N}(y)) | x \leq y\}, \tag{4.20}$$

which measures how much $\overline{N}$ is less than or equal to $\overline{M}$. We write $\overline{N} < \overline{M}$ if $v(\overline{N} \leq \overline{M}) = 1$ but $v(\overline{M} \leq \overline{N}) < \theta$, where θ is some fixed fraction in $(0, 1]$. Let us use $\theta = 0.8$ in this book. Then $\overline{N} < \overline{M}$ if $v(\overline{N} \leq \overline{M}) = 1$ and $v(\overline{M} \leq \overline{N}) < 0.8$. We define $\overline{M} \approx \overline{N}$ when both $\overline{M} < \overline{N}$ and $\overline{N} < \overline{M}$ are false. $\overline{M} \leq \overline{N}$ means $\overline{M} < \overline{N}$ or $\overline{M} \approx \overline{N}$.

Is this $\leq$ a total order? Is it a partial order? All we need to do is check and see if transitivity holds.

Example 4.5.2

For fuzzy numbers $\overline{M}$ and $\overline{N}$ let $\overline{M}[\alpha] = [m_1(\alpha), m_2(\alpha)]$ and $\overline{N}[\alpha] = [n_1(\alpha), n_2(\alpha)]$. We define $\overline{M} \leq \overline{N}$ if $m_1(\alpha) \leq n_1(\alpha)$ and $m_2(\alpha) \leq n_2(\alpha)$ for $0 \leq \alpha \leq 1$. Here we defined $\leq$ directly in terms of the $\alpha - cuts$ and not from $<$ and $\approx$. Therefore we need to check all four properties in order for $\leq$ to be a total order. Clearly, $\overline{M} \leq \overline{M}$ is true. Also, if $\overline{M} \leq \overline{N}$ and $\overline{N} \leq \overline{M}$, then we see that $m_1(\alpha) = n_1(\alpha)$ and $m_2(\alpha) = n_2(\alpha)$ for all α and $\overline{M} = \overline{N}$ which is property #3. Now check out the other two properties (#2, #4) to see if they are also true.

Example 4.5.3

This method is called the defuzzification method because we: (1) first assign a real number to each fuzzy number; and (2) then use the natural total ordering of the real numbers to get a total ordering for the fuzzy numbers. Defuzzification is discussed in more detail in the next section.

To do this method we need a function ψ mapping fuzzy numbers into the real numbers. Let $\psi(\overline{M}) = m$, $\overline{M}$ a fuzzy number and m a real number. Then $\overline{M} \leq \overline{N}$ if $\psi(\overline{M}) = m \leq n = \psi(\overline{N})$. This will always give a total ordering on the set of fuzzy numbers.

As an example of such a function ψ let $\psi(\overline{M})$ = the midpoint of the core of $\overline{M}$. Now the core of $\overline{M}$ is $\overline{M}[1] = [b, c]$ some closed interval. Then $\psi(\overline{M}) = (b + c)/2$. We could have $b = c$, the core a single point, and then $\psi(\overline{M}) = b$. Let us apply this method to $\overline{M} = (2/3, 5/6)$ and $\overline{N} = (0/4.5/5)$ shown in Figure 4.11. We compute $\psi(\overline{M}) = 4$ and $\psi(\overline{N}) = 4.5$ and conclude that $\overline{M} < \overline{N}$. But from Figure 4.15 many of us would disagree with this conclusion. This situation is common in the ordering of fuzzy numbers. Whatever $\leq$ you choose to use , some example like in Figure 4.15 can be constructed, where you do not agree with the result $\overline{M} < \overline{N}$.

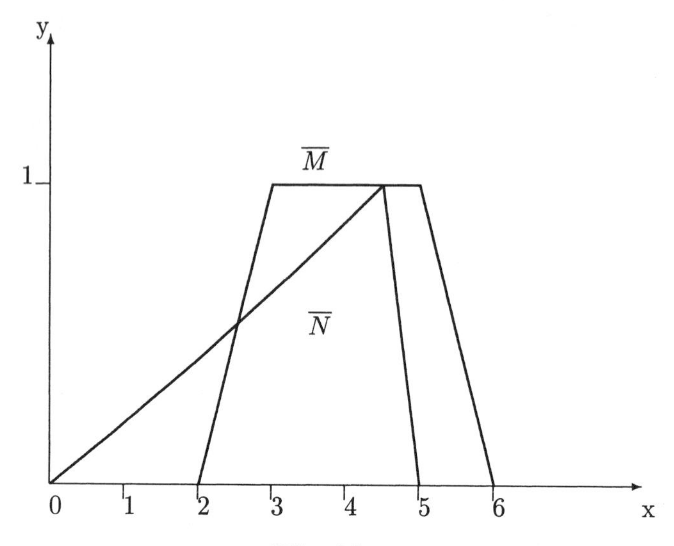

Figure 4.15: $\overline{M} < \overline{N}$ in Example 4.5.3

Example 4.5.4

In this method we first compute measures (scores) for $\overline{N} < \overline{M}$, $\overline{N} \approx \overline{M}$ and $\overline{N} > \overline{M}$ and choose the result that has the highest score. $\overline{M}$ will be a fixed fuzzy number and we compare fuzzy number $\overline{N}$ to $\overline{M}$.

Fuzzy sets

$$\overline{M}_u(x) = \sup\{\overline{M}(y) | y \leq x\}, \tag{4.21}$$

and

$$\overline{M}_l(x) = \sup\{\overline{M}(y) | y \geq x\}, \tag{4.22}$$

are first constructed. For triangular $\overline{M}$, Figure 4.16 shows $\overline{M}_l$, $1 - \overline{M}_l$, and Figure 4.17 shows $\overline{M}_u$ and $1 - \overline{M}_u$.

Then we make the comparisons

$$v(\overline{N} < \overline{M}) = \sup\{\min(1 - \overline{M}_u(x), \overline{N}(x)) | \textit{ all real } x\}, \tag{4.23}$$

$$v(\overline{M} \approx \overline{N}) = \sup\{\min(\overline{M}(x), \overline{N}(x)) | \textit{ all real } x\}, \tag{4.24}$$

and

$$v(\overline{N} > \overline{M}) = \sup\{\min(1 - \overline{M}_l(x), \overline{N}(x)) | \textit{ all real } x\}. \tag{4.25}$$

The fuzzy set "less than $\overline{M}$" is to be $1 - \overline{M}_u$, which is compared to $\overline{N}$ to get $v(\overline{N} < \overline{M})$. The height of the intersection of $\overline{M}$ and $\overline{N}$ is $v(\overline{M} \approx \overline{N})$. The fuzzy set "greater than $\overline{M}$" is $1 - \overline{M}_l$ which is compared to $\overline{N}$ for the value $v(\overline{N} > \overline{M})$. The largest of these three numbers gives the result. There can be ties, or there are two values which are equal. For example (no ties), if $v(\overline{N} < \overline{M}) = 0.25$, $v(\overline{N} \approx \overline{M}) = 0.75$ and $v(\overline{N} > \overline{M}) = 0.67$, then we conclude that $\overline{N} \approx \overline{M}$. As before, $\leq$ means $<$ or $\approx$.

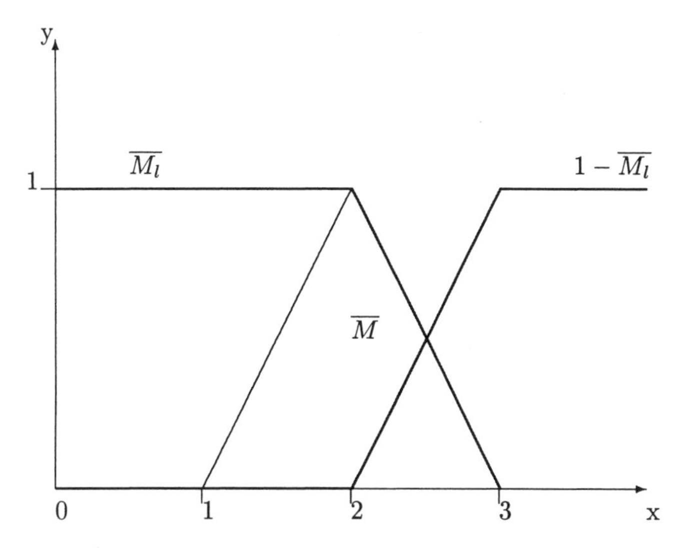

Figure 4.16: $\overline{M}_l$ and $1 - \overline{M}_l$ for Example 4.5.4

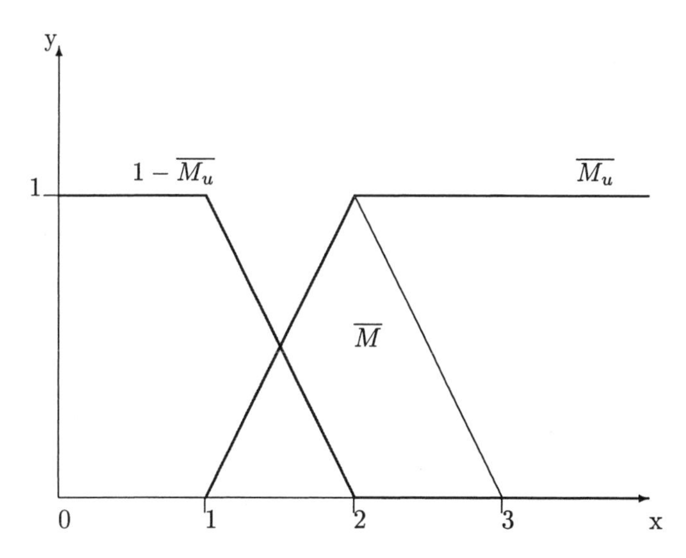

Figure 4.17: $\overline{M}_u$ and $(1 - \overline{M}_u)$ in Example 4.5.4

Clearly, $\overline{M} \leq \overline{M}$ is true by this method since we would get $v(\overline{M} \approx \overline{M}) = 1$ and $v(\overline{M} < \overline{M}) < 1$. Also, it looks like given any two fuzzy numbers $\overline{M}$ and $\overline{N}$ we will get $\overline{N} \leq \overline{N}$ or $\overline{M} \leq \overline{N}$. But does this method give a partial ordering? Does it give a total ordering?

4.5.1 Exercises

1. Show that if $\approx$ is not transitive, then $\leq$, defined as $<$ or $\approx$, is also not transitive.

2. In Example 4.5.1 is the $\leq$ defined there transitive? Hint: show that $\approx$ is not transitive.

3. Is the ordering defined in Example 4.5.2 transitive? Does it satisfy property # 2?

4. Come up with 2 or 3 more defuzzification methods (Example 4.5.3) for a total ordering the set of fuzzy numbers.

5. In Example 4.5.4 does $\leq$ satisfy properties #2 (transitive) and #3?

6. In Example 4.5.4, let $a = v(\overline{N} < \overline{M})$, $b = v(\overline{N} \approx \overline{M})$ and $c = v(\overline{N} > \overline{M})$. If the following result is possible, then determine the final decision on comparing $\overline{N}$ and $\overline{M}$ if:

   a. $a = b > c$,

   b. $a < b = c$,

   c. $a = b = c$, and

   d. $a = c > b$.

7. Show that if we first define "$<$" between fuzzy numbers and then say $\overline{M} \approx \overline{N}$ if both $\overline{M} < \overline{N}$ and $\overline{M} > \overline{N}$ are false, then properties #1, #3 and #4 automatically hold.

8. Define $\overline{M} \leq \overline{N}$ if and only if $\overline{M}(x) \leq \overline{N}(x)$ for all x. Is this $\leq$ a partial order?

9. How would you define $\leq$ between discrete fuzzy subsets of the real numbers?

10. Define $\overline{M} \leq \overline{N}$ if and only if $\overline{\min}(\overline{M}, \overline{N}) = \overline{M}$. Is this $\leq$ a partial order?

11. Define $\overline{M} \leq \overline{N}$ if and only if $\overline{\max}(\overline{M}, \overline{N}) = \overline{N}$. Is this $\leq$ a partial order?

12. Let $\overline{\max}(\overline{M}, \overline{N}) = \overline{P}$. Write $\overline{M} < \overline{N}$ if $D(\overline{N}, \overline{P}) < D(\overline{M}, \overline{P})$ where D is the distance measure (metric) given in equation (3.80) of Section 3.7. Then write $\overline{M} \approx \overline{N}$ if both $\overline{M} < \overline{N}$ and $\overline{N} < \overline{M}$ are false. Is this $\leq$ a partial order?

13. Suppose $\leq$ is the total ordering obtained from the center of the core defuzzification method in Example 4.5.3. Let $\overline{M}, \overline{N}, \overline{P}, \overline{Q}$ be continuous fuzzy numbers. Also assume that $\overline{M} \leq \overline{N}$. Determine if the following equations are always true:

a. If $\overline{P} > 0$, then $\overline{M}\,\overline{P} \leq \overline{N}\,\overline{P}$,

b. If $\overline{P} < 0$, then $\overline{M}\,\overline{P} \geq \overline{N}\,\overline{P}$,

c. $\overline{M} + \overline{Q} \leq \overline{N} + \overline{Q}$,

d. $\overline{M} - \overline{Q} \leq \overline{N} - \overline{Q}$,

e. Are any of these results true for any total ordering?

14. For two intervals $I = [a, b]$ and $J = [c, d]$ we define $I \leq J$ if and only if $b \leq c$. Determine if the following definitions of "$\leq$ " give a partial (total) ordering on the set of fuzzy numbers:

    a. $\overline{M} \leq \overline{N}$ if and only if $base(\overline{M}) \leq base(\overline{N})$, and

    b. $\overline{M} \leq \overline{N}$ if and only if $core(\overline{M}) \leq core(\overline{N})$,

15. Let "$\leq$ " be a partial order on the set of fuzzy numbers. Define $X = \{(\overline{M}, \overline{N}) | \overline{M}, \overline{N}$ fuzzy numbers$\}$. Next define $\preceq$ on X as follows: $(\overline{M}, \overline{N}) \preceq (\overline{P}, \overline{Q})$ if and only if: (1) $\overline{M} \leq \overline{P}$; (2) if $\overline{M} = \overline{P}$, then $\overline{N} \leq \overline{Q}$. Is " $\preceq$ " a partial order on X?

4.6 Defuzzification

We mentioned defuzzification in the previous section as a method of assigning a real number to a fuzzy subset of the real numbers. In this section we will look at some methods of defuzzification. The operation of defuzzification is a function, which we now call "*defuzz*", mapping fuzzy subsets of the real numbers into the real numbers. We will restrict the discussion to continuous fuzzy subsets or to discrete fuzzy subsets of the reals.

Defizzification is very important in the fuzzy controller. We do not discuss the fuzzy controller in this book , but its internal method of processing information is very similar to the fuzzy reasoning methods discussed in Chapter 14. In both systems the final conclusion turns out to be a fuzzy subset of the reals, like the one in Figure 14.3. If this final conclusion is to be communicated to a machine (set new speed, new voltage, etc.) it must be defuzzified because a machine will not understand a complete fuzzy set. So if the final conclusion is fuzzy set $\overline{B}$, then $defuzz(\overline{B})$ will be sent to the machine.

Let $\overline{A} = (a_1/a_2/a_3)$ where $a_3 - a_2 = a_2 - a_1$, or $\overline{A}$ is a symmetric triangular fuzzy number. Then many would agree that $defuzz(\overline{A}) = a_2$ is a reasonable defuzzification. But what if $\overline{A}$ is not symmetric? We now present some popular defuzzification methods through the following examples.

Example 4.6.1

The centroid defuzzifier. Let $\overline{A} = (-2/1, 2/6)$. The centroid defuzzifier $\delta = defuzz(\overline{A})$ is

$$\delta = \int_{-2}^{6} (x\overline{A}(x))dx \div (area), \tag{4.26}$$

where "*area*" is the area of the trapezoidal fuzzy number. To evaluate δ we need the functions $y = f_1(x)$ ($f_2(x)$) for the left (right) side of $\overline{A}$. Now $\overline{A}[\alpha] = [-2 + 3\alpha, 6 - 4\alpha]$, so we find that $y = f_1(x) = (x + 2)/3$ and $y = f_2(x) = (6 - x)/4$. Then the numerator of δ is

$$\int_{-2}^{1} xf_1(x)dx + \int_{1}^{2} xdx + \int_{2}^{6} xf_2(x)dx, \tag{4.27}$$

which equals

$$0 + 3/2 + 20/3. \tag{4.28}$$

The area is easily found to be 4.50. So, $\delta = 49/27$.

Example 4.6.2

The centroid defuzzifier for discrete fuzzy sets. Suppose $\overline{A}$ is a discrete fuzzy subset of the real numbers given by

$$\overline{A} = \{\frac{\mu_1}{x_1}, \cdots, \frac{\mu_n}{x_n}\}, \tag{4.29}$$

for x_i real numbers and μ_i in $[0, 1]$, $1 \leq i \leq n$. Then if $\delta = defuzz(\overline{A})$, we compute δ as

$$\delta = \frac{\sum_{i=1}^{n} x_i \mu_i}{\sum_{i=1}^{n} \mu_i}. \tag{4.30}$$

Example 4.6.3

The center of maxima defuzzifier. Let

$$C_m = \{x | \overline{A}(x) = ht(\overline{A})\}, \tag{4.31}$$

where $ht(\overline{A})$ = height of $\overline{A}$ (which is one for fuzzy numbers). Then

$$defuzz(\overline{A}) = (\min C_m + maxC_m)/2. \tag{4.32}$$

If $\overline{A} \approx (a_1/a_2/a_3)$, then $defuzz(\overline{A}) = a_2$ since $C_m = \{a_2\}$. If $\overline{A} \approx (a_1/a_2, a_3/a_4)$, then $defuzz(\overline{A}) = (a_2 + a_3)/2$ because $C_m = [a_2, a_3]$ which is the core of $\overline{A}$. For discrete $\overline{A}$

$$\overline{A} = \{\frac{0}{0}, \frac{0.8}{1}, \frac{0.6}{2}, \frac{0.8}{3}, \frac{0.3}{4}, \frac{0.1}{5}\}, \tag{4.33}$$

we have $defuzz(\overline{A})$ = (1+3)/2 . Sometimes, for discrete fuzzy sets $defuzz(\overline{A})$ does not equal a member of X. When this happens you might round up, or round down, to get $defuzz(\overline{A})$ equal to a member of X. For example if

$$\overline{A} = \{\frac{0.5}{0}, \frac{0.7}{1}, \frac{0.3}{2}, \frac{0.6}{3}, \frac{0.7}{4}, \frac{0}{5}\}, \tag{4.34}$$

then $defuzz(\overline{A}) = (1 + 4)/2 = 2.5$. Then we could round to 3 or to 2.

Example 4.6.4

Mean of the maxima defuzzifier. For a discrete fuzzy set $\overline{A}$ define C_m as in Example 4.6.3. Then

$$defuzz(\overline{A}) = \sum_{i=1}^{K} \{x_i | x_i \in C_m\}/K, \tag{4.35}$$

where K is the total number of points in the set C_m.

4.6.1 Exercises

1. Compute the centroid defuzzifier for $\overline{A} = (-3/-1/4)$.

2. Compute the centroid defuzzifier for

$$\overline{A} = \{\frac{0.9}{-3}, \frac{0.7}{-2}, \frac{0}{-1}, \frac{0.3}{0}, \frac{0.3}{1}, \frac{1}{2}, \frac{0.2}{3}\}.$$

3. Extend the mean of the maxima defuzzifier (Example 4.6.4) to continuous fuzzy subsets of the real numbers, not necessarily fuzzy numbers.

4. Compute the mean of the maxima defuzzifier for the two discrete fuzzy sets in Example 4.6.3.

5. Given a "$\leq$" between continuous fuzzy numbers and if $\overline{M} \leq \overline{N}$, is $defuzz(\overline{M}) \leq defuzz(\overline{N})$? Use the centroid defuzzifier and the center of maxima defuzzifier. Answer the question for the $\leq$ given in:"

   a. Example 4.5.1,

   b. Example 4.5.2, and

   c. Example 4.5.3.

6. Given two continuous fuzzy numbers $\overline{M}$ and $\overline{N}$ is $defuzz(\overline{M} + \overline{N}) = defuzz(\overline{M}) + defuzz(\overline{N})$? Use t-norm $T = \min$ for the sum of fuzzy numbers. Also use the centroid defuzzifier and the center of maxima defuzzifier.

7. Given continuous fuzzy number $\overline{M}$ and real number r, do we get $defuzz(\overline{M} + r) = defuzz(\overline{M}) + r$? Use t-norm min for the addition of the fuzzy number and the real number. Also, use the centroid defuzzifier and the center of maxima defuzzifier.

8. Given two continuous fuzzy numbers $\overline{M}$ and $\overline{N}$ determine if the following equation is true or false. Use t-norm min for $\cap$ and t-conorm max for $\cup$ and the centroid defuzzifier:

   a. $defuzz(\overline{M} \cap \overline{N}) \leq defuzz(\overline{M})$ [and $\leq defuzz(\overline{N})$], and

   b. $defuzz(\overline{M} \cup \overline{N}) \geq defuzz(\overline{M})$ [and $\geq defuzz(\overline{N})$].

9. If we define $D(\overline{M}, \overline{N})$, the distance between two continuous fuzzy numbers $\overline{M}$ and $\overline{N}$, as

$$D(\overline{M}, \overline{N}) = |defuzz(\overline{M}) - defuzz(\overline{N})|,$$

   is D a metric (see section 3.7)? Can you define a method of defuzzification so that D is a metric?

10. Is

$$defuzz(\overline{max}(\overline{M},\overline{N})) \geq defuzz(\overline{M}), [\geq defuzz(\overline{N})],$$

true for continuous fuzzy numbers $\overline{M}$ and $\overline{N}$? Use the centroid defuzzifier.

11. Is

$$defuzz(\overline{\min}(\overline{M},\overline{N})) \leq defuzz(\overline{M}), [\leq defuzz(\overline{N})],$$

true for continuous fuzzy numbers $\overline{M}$ and $\overline{N}$? Use the centroid defuzzifier.

12. How would you defuzzify a level 2 and a type 2 fuzzy set?

Chapter 5

Fuzzy Equations

5.1 Introduction

In this chapter we start solving fuzzy equations. In the next section we look at solutions to the simple fuzzy linear equation. However, the problems involved in solving this simple fuzzy linear equation are sufficiently complicated to indicate three possible solution techniques: the classical method; the extension principle procedure; and the α-cut and interval arithmetic method. All three solutions are illustrated on the fuzzy linear equation and these three methods will be used in the rest of the book. In the third section we then apply these solutions to the fuzzy quadratic equation.

5.2 Linear Equations

In algebra one of the first things you do is solve linear equations. If

$$ax + b = c, \tag{5.1}$$

for given values of a, b, c with $a \neq 0$, you are to solve for x. The solution is $x = (c - b)/a$. But to get this solution you first subtracted b from both sides of the equation and then you multiplied both sides by $(1/a)$.

In fuzzy algebra the first thing we want to do is solve

$$\overline{A} \cdot \overline{X} + \overline{B} = \overline{C}, \tag{5.2}$$

for $\overline{X}$ given triangular fuzzy numbers $\overline{A}$, $\overline{B}$, $\overline{C}$. Let $\overline{A} = (a_1/a_2/a_3)$, $\overline{B} = (b_1/b_2/b_3)$ and $\overline{C} = (c_1/c_2/c_3)$ and we assume zero does not belong to $\overline{A}[0] = [a_1, a_3]$. Assuming $\overline{X}$ can be a triangular, or a triangular shaped fuzzy number, we let $\overline{X} \approx (x_1/x_2/x_3)$.

Following the solution to the crisp equation (5.1) we obtain

$$(1/\overline{A})(\overline{A} \cdot \overline{X} + \overline{B} - \overline{B}) = (\overline{C} - \overline{B})/\overline{A}. \tag{5.3}$$

But the left hand side of equation (5.3) does not equal $\overline{X}$ since $\overline{B} - \overline{B} \neq 0$ and $\overline{A}/\overline{A} \neq 1$. For example, if $\overline{B} = (1/2/3)$, then $\overline{B} - \overline{B} = (-2/0/2)$ not zero. Also, if $\overline{A} = (1/2/3)$, then $\overline{A}/\overline{A}$ is a triangular shaped fuzzy number $\approx (\frac{1}{3}/1/3)$ not equal to one.

This shows a major problem in solving fuzzy equations: some basic operations we use to solve crisp equations do not hold for fuzzy equations. As a result of this problem several solution concepts have been introduced for fuzzy equations. We will now present three of these solutions.

5.2.1 Classical Solution

The classical solution, written $\overline{X}_c$ if it exists, involves substitution of α-cuts of $\overline{A}$, $\overline{B}$, $\overline{C}$ and $\overline{X}_c$ into equation (5.2), and using interval arithmetic, solve for the α-cuts of $\overline{X}_c$. Let $\overline{A}[\alpha] = [a_1(\alpha), a_2(\alpha)]$, $\overline{B}[\alpha] = [b_1(\alpha), b_2(\alpha)]$, $\overline{C}[\alpha] = [c_1(\alpha), c_2(\alpha)]$ and $\overline{X}_c[\alpha] = [x_1(\alpha), x_2(\alpha)]$. Substitute these into equation (5.2) giving

$$[a_1(\alpha), a_2(\alpha)][x_1(\alpha), x_2(\alpha)] + [b_1(\alpha), b_2(\alpha)] = [c_1(\alpha), c_2(\alpha)]. \tag{5.4}$$

Now use interval arithmetic to solve for $x_1(\alpha)$ and $x_2(\alpha)$. We say this method defines a solution $\overline{X}_c$ if $[x_1(\alpha), x_2(\alpha)]$ defines the α-cuts of a fuzzy number. For this to be true we need:

1. $x_1(\alpha)$ to be a monotonically increasing function of α, $0 \leq \alpha \leq 1$;
2. $x_2(\alpha)$ to be a monotonically decreasing function of α, $0 \leq \alpha \leq 1$; and
3. $x_1(1) \leq x_2(1)$.

Since $\overline{A}$, $\overline{B}$, $\overline{C}$ are all continuous fuzzy numbers the $x_i(\alpha)$ will be continuous functions of α. If $x_1(1) < x_2(1)$ we get $\overline{X}_c$ a trapezoidal shaped fuzzy number.

Example 5.2.1.1

Let $\overline{A} = (1/2/3)$, $\overline{B} = (-3/-2/-1)$ and $\overline{C} = (3/4/5)$. Then $\overline{A}[\alpha] = [1+\alpha, 3-\alpha]$, $\overline{B}[\alpha] = [-3+\alpha, -1-\alpha]$, $\overline{C}[\alpha] = [3+\alpha, 5-\alpha]$. Since $\overline{A} > 0$, $\overline{B} < 0$ and $\overline{C} > 0$ we must have $\overline{X}_c > 0$. From interval arithmetic equation (5.4) becomes

$$[a_1(\alpha)x_1(\alpha) + b_1(\alpha), a_2(\alpha)x_2(\alpha) + b_2(\alpha)] = [c_1(\alpha), c_2(\alpha)], \tag{5.5}$$

or

$$x_1(\alpha) = \frac{6}{1+\alpha}, \tag{5.6}$$

$$x_2(\alpha) = \frac{6}{3-\alpha}, \tag{5.7}$$

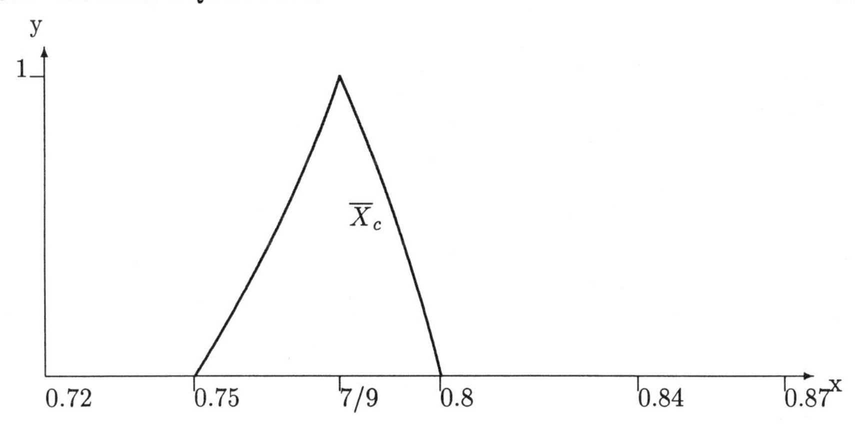

Figure 5.1: Solution in Example 5.2.1.2

after substituting for $a_1(\alpha)$,...,$c_2(\alpha)$ and solving for the $x_i(\alpha)$.

But $x_1(\alpha)$ is a decreasing function of α and $x_2(\alpha)$ is an increasing function of α. Hence, $\overline{X}_c$ does not exist.

Example 5.2.1.2

Now let $\overline{A} = (8/9/10)$, so that $\overline{A}[\alpha] = [8+\alpha, 10-\alpha]$, $\overline{B} = (-3/-2/-1)$ as in Example 5.2.1.1 and $\overline{C} = (3/5/7)$ so that $\overline{C}[\alpha] = [3+2\alpha, 7-2\alpha]$. Again $\overline{X}_c > 0$ and solving we obtain

$$x_1(\alpha) = \frac{6+\alpha}{8+\alpha}, \tag{5.8}$$

$$x_2(\alpha) = \frac{8-\alpha}{10-\alpha}, \tag{5.9}$$

We see that $x_1(\alpha)$ is an increasing function of α (its derivative is positive), $x_2(\alpha)$ is a decreasing function of α (its derivative is negative) and $x_1(1) = 7/9 = x_2(1)$. The solution $\overline{X}_c$ exists and is shown in Figure 5.1.

Working more examples like Examples 5.2.1.1 and 5.2.1.2 we conclude too often the fuzzy linear equation has no $\overline{X}_c$ solution. This motivates us to consider other types of solution.

5.2.2 Extension Principle Solution

We continue to look for solutions to the fuzzy linear equation in equation (5.2). This solution fuzzifies the crisp solution. The crisp solution is $x = (c-b)/a$ and we fuzzify it by substituting $\overline{A}$ for a, $\overline{B}$ for b and $\overline{C}$ for c. The result is

$$\overline{X} = (\overline{C} - \overline{B})/\overline{A}. \tag{5.10}$$

There are two basic ways to evaluate equation (5.10): (1) using the extension principle; or (2) using α-cuts and interval arithmetic. The extension principle procedure gives $\overline{X}_e$ and the α-cut and interval arithmetic method produces $\overline{X}_i$. In this section we look at $\overline{X}_e$ and $\overline{X}_i$ is in the next section.

We have mentioned the extension principle before (Sections 4.3.1 and 4.4) and it will be used many times in this book. So, now let us formally present the extension principle. This principle is used to fuzzify equations and functions. Let $y = f(x_1, \cdots, x_n)$ be a function of n real variables x_i. Substitute continuous fuzzy numbers $\overline{A}_i$ for $x_i, 1 \leq i \leq n$, giving

$$\overline{Y} = f(\overline{A}_1, \cdots, \overline{A}_n). \tag{5.11}$$

The extension principle is used to obtain the membership function for $\overline{Y}$. First define

$$\pi(x_1, \cdots, x_n) = \min_{1 \leq i \leq n} \overline{A}_i(x_i). \tag{5.12}$$

and then

$$\overline{Y}(y) = \sup\{\pi(x_1, \cdots, x_n) | f(x_1, \cdots, x_n) = y\}. \tag{5.13}$$

Apply this to equation (5.10) giving

$$\overline{X}_e(x) = \sup\{\pi(a, b, c) | (c - b)/a = x\}, \tag{5.14}$$

where

$$\pi(a, b, c) = \min(\overline{A}(a), \overline{B}(b), \overline{C}(c)). \tag{5.15}$$

In general, this computation, equations (5.14) and (5.15), looks difficult but for continuous functions, the f in equation (5.11), we may find the α-cuts of $\overline{X}_e$ as follows:

$$x_1(\alpha) = \min\{(c - b)/a | a \in \overline{A}[\alpha], b \in \overline{B}[\alpha], c \in \overline{C}[\alpha]\}, \tag{5.16}$$

$$x_2(\alpha) = \max\{(c - b)/a | a \in \overline{A}[\alpha], b \in \overline{B}[\alpha], c \in \overline{C}[\alpha]\}, \tag{5.17}$$

where $\overline{X}_e[\alpha] = [x_1(\alpha), x_2(\alpha)]$. $\overline{X}_e$ always exists and is a triangular shaped fuzzy number.

Example 5.2.2.1

This continues Example 5.2.1.1. We need to evaluate equations (5.16) and (5.17). This is easy because $(c - b)/a$ is: (1) an increasing function of c; and (2) a decreasing function of a and b. So

$$x_1(\alpha) = \frac{c_1(\alpha) - b_2(\alpha)}{a_2(\alpha)} = \frac{4 + 2\alpha}{3 - \alpha}, \tag{5.18}$$

$$x_2(\alpha) = \frac{c_2(\alpha) - b_1(\alpha)}{a_1(\alpha)} = \frac{8 - 2\alpha}{1 + \alpha}. \tag{5.19}$$

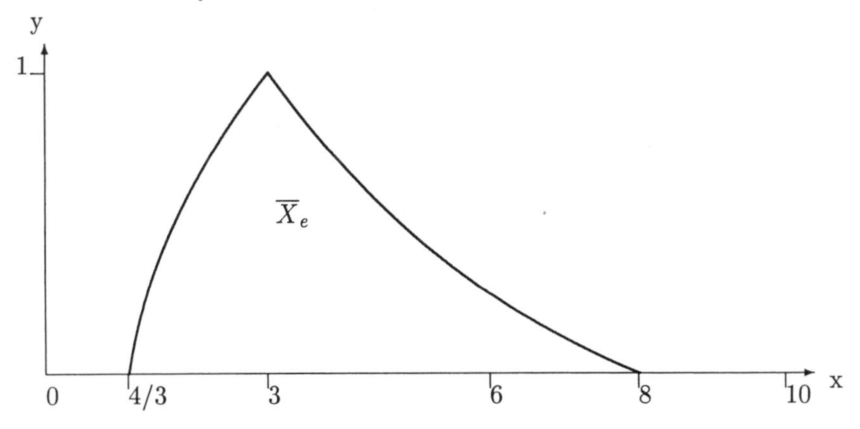

Figure 5.2: Solution to Example 5.2.2.1

$\overline{X}_e$ is shown in Figure 5.2.

If you substitute $\overline{X}_e$ back into the fuzzy linear equation

$$\overline{A} \cdot \overline{X}_e + \overline{B} = \overline{C} \tag{5.20}$$

it may, or may not, be true. That is, the extension principle solution may, or may not, satisfy the original equation. We would evaluate equation (5.20) using α-cuts and interval arithmetic. Equation (5.20) is not true for the $\overline{X}_e$ of Example 5.2.2.1.

5.2.3 Alfa-Cut and Interval Arithmetic Solution

$\overline{X}_i$ is obtained from equation (5.10) by using α-cuts and interval arithmetic. So, if $\overline{X}_i[\alpha] = [x_1(\alpha), x_2(\alpha)]$, then

$$[x_1(\alpha), x_2(\alpha)] = \frac{[c_1(\alpha), c_2(\alpha)] - [b_1(\alpha), b_2(\alpha)]}{[a_1(\alpha), a_2(\alpha)]}, \tag{5.21}$$

or

$$x_1(\alpha) = \frac{c_1(\alpha) - b_2(\alpha)}{a_2(\alpha)}, \tag{5.22}$$

$$x_2(\alpha) = \frac{c_2(\alpha) - b_1(\alpha)}{a_1(\alpha)}, \tag{5.23}$$

if $a_1(\alpha) > 0$ all α. So, if $\overline{A} > 0$ we get the same result as for $\overline{X}_e$. That is, $\overline{X}_i = \overline{X}_e$ if $\overline{A} > 0$.

In general, $\overline{X}_e$ and $\overline{X}_i$ always exist but they may not satisfy the original fuzzy linear equation.

Example 5.2.3.1

This continues Example 5.2.1.2. As in Example 5.2.2.1 we get for $\overline{X}_i$

$$x_1(\alpha) = \frac{c_1(\alpha) - b_2(\alpha)}{a_2(\alpha)} = \frac{4 + 3\alpha}{10 - \alpha}, \tag{5.24}$$

$$x_2(\alpha) = \frac{c_2(\alpha) - b_1(\alpha)}{a_1(\alpha)} = \frac{10 - 3\alpha}{8 + \alpha}. \tag{5.25}$$

We also see that $\overline{X}_i = \overline{X}_e$.

In this example $\overline{X}_c$ exists and notice that $\overline{X}_c$ is "inside" $\overline{X}_e$, or $\overline{X}_c(x) \leq \overline{X}_e(x)$ all x.

5.2.4 Exercises

1. Let $\overline{A} = (-3/-2/-1)$, $\overline{B} = (-3/-2/-1)$ and $\overline{C} = (3/4/5)$. Assume $\overline{X}_c < 0$.

   a. Find $\overline{X}_c$, if it exists.

   b. Find $\overline{X}_e$. Does $\overline{X}_e$ satisfy the equation?

   c. Find $\overline{X}_i$. Does $\overline{X}_i = \overline{X}_e$?

2. Show if $\overline{X}_c$ exists, then $\overline{X}_c(x) \leq \overline{X}_e(x)$ all x.

3. Let $\overline{A} = (1/2/3)$, $\overline{B} = (-3/-2/-1)$, $\overline{C} = (-5/-4/-3)$. Assume $\overline{X}_c < 0$.

   a. Find $\overline{X}_c$, if it exists.

   b. Find $\overline{X}_e$. Does $\overline{X}_e$ satisfy the equation?

   c. Find $\overline{X}_i$. Does $\overline{X}_i = \overline{X}_e$?

4. Is it true that $\overline{X}_i = \overline{X}_e$ for all $\overline{A}$, $\overline{B}$, $\overline{C}$?

5. Is it true that $\overline{X}_e(x) \leq \overline{X}_i(x)$ all x, for all $\overline{A}$, $\overline{B}$, $\overline{C}$?

6. Is it true that $\overline{X}_c$, if it exists, always satisfies the original equation $\overline{A} \cdot \overline{X}_c + \overline{B} = \overline{C}$?

7. Show that $\overline{A} \cdot \overline{X}_e + \overline{B} \neq \overline{C}$ for $\overline{X}_e$ in Example 5.2.2.1.

8. Does $\overline{A} \cdot \overline{X}_i + \overline{B} = \overline{C}$ for $\overline{X}_i$ in Example 5.2.3.1?

9. In the following problems solve for $\overline{X}_c$ (if it exists), $\overline{X}_e$ and $\overline{X}_i$. Show $\overline{X}_e \leq \overline{X}_i$ and $\overline{X}_c \leq \overline{X}_e$ if $\overline{X}_c$ exists. Check to see if $\overline{A} \cdot \overline{X}_e + \overline{B} = \overline{C}$ and $\overline{A} \cdot \overline{X}_i + \overline{B} = \overline{C}$. Let $\overline{U} = (1/2/3)$, $\overline{V} = (3/4/5)$.

   a. $\overline{A} = \overline{U}$, $\overline{B} = \overline{U}$, $\overline{C} = -\overline{V}$.

   b. $\overline{A} = \overline{U}$, $\overline{B} = \overline{U}$, $\overline{C} = \overline{V}$.

   c. $\overline{A} = \overline{U}$, $\overline{B} = -\overline{U}$, $\overline{C} = -\overline{V}$.

   d. $\overline{A} = -\overline{U}$, $\overline{B} = \overline{U}$, $\overline{C} = \overline{V}$.

   e. $\overline{A} = -\overline{U}$, $\overline{B} = \overline{U}$, $\overline{C} = -\overline{V}$.

   f. $\overline{A} = -\overline{U}$, $\overline{B} = -\overline{U}$, $\overline{C} = -\overline{V}$.

10. Consider a fuzzy linear equation $\overline{A}\ \overline{X} + \overline{B} = \overline{C}$ where the classical solution $\overline{X}_c$ does not exist. Then we define a substitute classical solution $\overline{X}_s$ to be that triangular shaped fuzzy number X that solves the following minimization problem

$$\min[D(\overline{A}\ \overline{X} + \overline{B}, \overline{C})],$$

for D the metric for continuous fuzzy numbers given in equation (3.80) in Section 3.7. Apply this method to find $\overline{X}_s$ in Example 5.2.1.1.

5.3 Other Fuzzy Equations

Now we could go on to fuzzy polynomial equations, systems of fuzzy linear equations (Chapter 11), etc., but we will only consider the fuzzy quadratic equation in this section.

We have defined three types of solutions to the fuzzy linear equation in the previous section. As we start to study other fuzzy equations we should have a solution strategy. A solution strategy is to decide which of these solutions we prefer, assuming the classical solution exists, and which we do not prefer. Our solution strategy will be: (1) use the classical solution if it exists because it always satisfies the original fuzzy equation; (2) if the classical solution fails to exist, use the extension principle solution, which may or may not satisfy the fuzzy equation; and (3) if the classical solution does not exist and the extension principle solution is difficult to obtain (equations like (5.16), (5.17)), then use the α-cut and interval arithmetic solution, which uses only interval arithmetic, as an approximation to the extension principle solution.

The fuzzy quadratic equation is

$$\overline{A} \cdot \overline{X}^2 + \overline{B} \cdot \overline{X} + \overline{C} = \overline{D}, \tag{5.26}$$

for triangular fuzzy numbers $\overline{A}$, $\overline{B}$, $\overline{C}$, $\overline{D}$ and $\overline{X}$ will be a triangular shaped fuzzy number. Notice that we did not write

$$\overline{A} \cdot \overline{X}^2 + \overline{B} \cdot \overline{X} + \overline{C} = 0, \tag{5.27}$$

because we could never get the left hand side of equation (5.27) exactly equal to (crisp) zero.

Before discussing solutions to equation (5.26) let us look at the crisp equation

$$ax^2 + bx + c = d, \tag{5.28}$$

where now d can equal zero. We know that equation (5.28) can have one, or two real solutions for x. If we allow complex numbers, then equation (5.28) has two solutions (counting a root of multiplicity two twice). However, in this book we will not deal with fuzzy complex numbers. We will only use real fuzzy numbers. So the constraints on $\overline{A}$, $\overline{B}$, $\overline{C}$, $\overline{D}$ in equation (5.26) is that the solution for $\overline{X}$ is a real fuzzy number, otherwise there is no solution.

We first look for a classical solution. Let $\overline{A}[\alpha] = [a_1(\alpha), a_2(\alpha)]$, $\overline{B}[\alpha] = [b_1(\alpha), b_2(\alpha)]$, $\overline{C}[\alpha] = [c_1(\alpha), c_2(\alpha)]$, $\overline{D}[\alpha] = [d_1(\alpha), d_2(\alpha)]$, and $\overline{X}[\alpha] = [x_1(\alpha), x_2(\alpha)]$. Substitute these $\alpha - cuts$ into equation (5.26) and solve for $x_1(\alpha)$ and $x_2(\alpha)$. To solve for the $x_i(\alpha)$ we need to know if $\overline{A}$, $\overline{B}$ and $\overline{X}$ are positive or negative. Let us assume for now that $\overline{A} > 0$, $\overline{B} > 0$ and $\overline{X} > 0$. Then we get two equations

$$a_i(\alpha)(x_i(\alpha))^2 + b_i(\alpha)x_i(\alpha) + c_i(\alpha) = d_i(\alpha), \tag{5.29}$$

$i = 1, 2$.

For notation define

$$S_1(a,b,c,d) = [-b + \sqrt{b^2 - 4a(c-d)}]/2a, \tag{5.30}$$

$$S_2(a,b,c,d) = [-b - \sqrt{b^2 - 4a(c-d)}]/2a. \tag{5.31}$$

Let us solve for the largest fuzzy solution. Then if $\overline{X} = \overline{X}_c$, for the classical solution, it is the solution whose α-cuts are $[x_1(\alpha), x_2(\alpha)]$ and we get

$$x_i(\alpha) = S_1(a_i(\alpha), b_i(\alpha), c_i(\alpha), d_i(\alpha)), \tag{5.32}$$

$i = 1, 2$.

For $\overline{X}_c$ to exist, $[x_1(\alpha), x_2(\alpha)]$ must define α-cuts of a triangular shaped fuzzy number. That is we need $\partial x_1/\partial\alpha > 0$ and $\partial x_2/\partial\alpha < 0$ for $0 < \alpha < 1$ and $x_1(1) = x_2(1)$. But also equation (5.32) must produce real numbers so we also require that

$$(b_i(\alpha))^2 - 4a_i(\alpha)(c_i(\alpha) - d_i(\alpha)) \geq 0 \tag{5.33}$$

$i = 1, 2$, and $0 \leq \alpha \leq 1$.

If these conditions are met $\overline{X}_c$ exists and is called the classical solution.

There could be another classical solution using S_2. Naturally, if $\overline{A} < 0$ and/or $\overline{B} < 0$ we may get different results.

It will be very difficult to meet all of these conditions, so quite often $\overline{X}_c$ does not exist, and we go on to $\overline{X}_e$, the extension principle solution. It fuzzifies S_1 and S_2. Working with S_1 let the α-cuts of $\overline{X}_e$ be $[x_1(\alpha), x_2(\alpha)]$ and then

$$x_1(\alpha) = \min\{S_1(a,b,c,d) | a \in \overline{A}[\alpha], \cdots, d \in \overline{D}[\alpha]\}, \tag{5.34}$$

$$x_2(\alpha) = \max\{S_1(a,b,c,d) | a \in \overline{A}[\alpha], \cdots, d \in \overline{D}[\alpha]\}, \tag{5.35}$$

for $0 \leq \alpha \leq 1$.

If it is difficult to find this min and max in equations (5.34) and (5.35), respectively, we look at $\overline{X}_i$, the α-cut and interval arithmetic solution.

We compute $\overline{X}_i$ by substituting α-cuts of $\overline{A}$, $\overline{B}$, $\overline{C}$, $\overline{D}$ into S_1 (or S_2). Using S_1 let the α-cut of $\overline{X}_i$ be $[x_1(\alpha), x_2(\alpha)]$. Then we see that

$$x_1(\alpha) = [-b_2(\alpha) + \sqrt{(b_1(\alpha))^2 - 4a_2(\alpha)(c_2(\alpha) - d_1(\alpha))}]/2a_2(\alpha), \tag{5.36}$$

$$x_2(\alpha) = [-b_1(\alpha) + \sqrt{(b_2(\alpha))^2 - 4a_1(\alpha)(c_1(\alpha) - d_2(\alpha))}]/2a_1(\alpha), \tag{5.37}$$

$0 \leq \alpha \leq 1$.

$\overline{X}_e$ and $\overline{X}_i$ always exist, assuming all the square roots exist as real numbers, and we believe that $\overline{X}_e \leq \overline{X}_i$ (see the exercises). However, $\overline{X}_e$ and $\overline{X}_i$ may not satisfy equation (5.26). To check this substitute α-cuts of $\overline{X}_e$ ($\overline{X}_i$) into equation (5.26), simplify using interval arithmetic, and see if the resulting equation is true.

Example 5.3.1

Suppose an investment firm wishes to set aside around A dollars to be invested at interest rate r so that after one year they may withdraw approximately B dollars. And then after two years the amount that is left will accumulate to about C dollars. Given values of A, B, and C solve for r. However, A, B and C are not known exactly and this uncertainty will be modeled using triangular fuzzy numbers $\overline{A}$, $\overline{B}$ and $\overline{C}$. Hence $\overline{r}$ will be a triangular shaped fuzzy number.

After one year the amount will be $\overline{A}+\overline{A}\overline{r}$. Now withdraw $\overline{B}$ and we have $(\overline{A}-\overline{B})+\overline{A}\overline{r}$ to begin the second year. At the end of the second year we have

$$[(\overline{A}-\overline{B})+\overline{A}\overline{r}]+[(\overline{A}-\overline{B})+\overline{A}\overline{r}]\overline{r}. \tag{5.38}$$

For positive fuzzy numbers multiplication distributes over addition (Section 4.3.4) which means $(\overline{X}+\overline{Y})\overline{Z}=\overline{X}\cdot\overline{Z}+\overline{Y}\cdot\overline{Z}$ if $\overline{X}>0$, $\overline{Y}>0$, $\overline{Z}>0$. Now $\overline{r}$ is a fuzzy interest rate so it is a fuzzy subset of $[0,1]$. We know $\overline{B}$ will be smaller than $\overline{A}$ so assume $\overline{A}-\overline{B}>0$. Then equation (5.38) becomes

$$(\overline{A}-\overline{B})+\overline{A}\overline{r}+(\overline{A}-\overline{B})\overline{r}+\overline{A}(\overline{r})^2, \tag{5.39}$$

or

$$\overline{A}(\overline{r})^2+\overline{D}\overline{r}+\overline{E}, \tag{5.40}$$

for $\overline{D}=\overline{A}+(\overline{A}-\overline{B})=2\overline{A}-\overline{B}$, $\overline{E}=\overline{A}-\overline{B}$.

So we must solve

$$\overline{A}(\overline{r})^2+\overline{D}\overline{r}+\overline{E}=\overline{C}, \tag{5.41}$$

for $\overline{r}$.

So now let us put some numbers in for $\overline{A}$, $\overline{B}$ and $\overline{C}$ and solve for $\overline{r}$. Assume $\overline{A}=(0.8/1.0/1.2)$, or approximate one million dollars, $\overline{B}=(0.20/0.25/0.30)$, or around \$250,000 and $\overline{C}=(0.60/0.90/1.20)$ about \$900,000. The unit $1=$ one million. Then $\overline{A}[\alpha]=[0.8+0.2\alpha,1.2-0.2\alpha]$, $\overline{B}[\alpha]=[0.20+0.05\alpha,0.30-0.05\alpha]$, $\overline{C}[\alpha]=[0.6+0.3\alpha,1.2-0.3\alpha]$ and $\overline{r}[\alpha]=[r_1(\alpha),r_2(\alpha)]$. Then $\overline{D}[\alpha]=[1.3+0.45\alpha,2.2-0.45\alpha]$ and $\overline{E}[\alpha]=[0.5+0.25\alpha,1-0.25\alpha]$. Substitute these α-cuts into equation (5.41) and solve for $r_i(\alpha)$, $i=1,2$. We are seeing if the classical solution exists. The equation for $r_1(\alpha)$ is

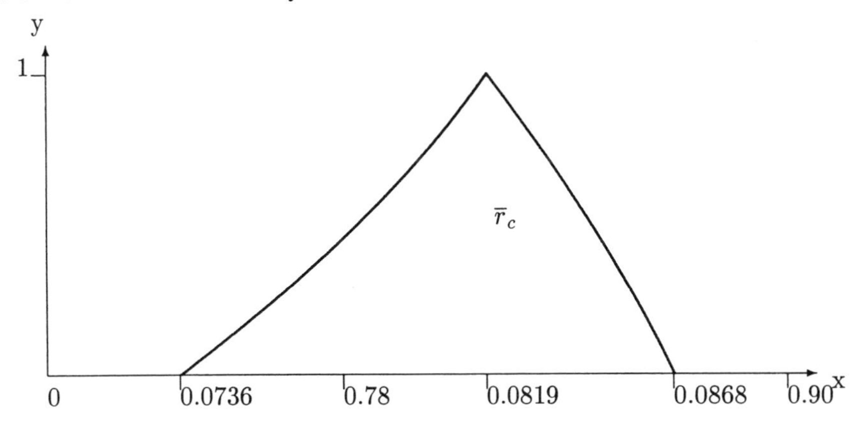

Figure 5.3: Fuzzy Interest Rate in Example 5.3.1

$$(0.8 + 0.2\alpha)(r_1(\alpha))^2 + (1.3 + 0.45\alpha)r_1(\alpha) + (-0.1 - 0.05\alpha) = 0, \quad (5.42)$$

so that

$$r_1(\alpha) = \frac{-(1.3 + 0.45\alpha) + \sqrt{M(\alpha)}}{2(0.8 + 0.2\alpha)}, \quad (5.43)$$

for

$$M(\alpha) = (1.3 + 0.45\alpha)^2 + 4(0.8 + 0.2\alpha)(0.1 + 0.05\alpha). \quad (5.44)$$

Similarly, we obtain $r_2(\alpha)$. We use the positive square root since $\overline{r}$ is in $[0, 1]$. We easily see that $M(\alpha) > 0$ all α, and the same result for $r_2(\alpha)$. Next we check $\partial r_1/\partial\alpha > 0$, $\partial r_2/\partial\alpha < 0$ which are true and also $r_1(1) = r_2(1) = 0.0819$. The classical solution exists and its graph is in Figure 5.3. Since the classical solution exists we do not calculate $\overline{r}_e$ or $\overline{r}_i$.

5.3.1 Exercises

1. We stated that $\overline{X}_i$ will be an approximation of $\overline{X}_e$. For the fuzzy quadratic equation show $\overline{X}_e(x) \leq \overline{X}_i(x)$ all x.

2. Explain why equation (5.27) can not have a solution for $\overline{X}$ if $\overline{A}$, $\overline{B}$, $\overline{C}$ are triangular fuzzy numbers.

3. Explain what changes need to be made in solving for $\overline{X}_c$, equations (5.29)–(5.32), if :

   a. $\overline{A} < 0$, $\overline{X} > 0$, $\overline{B} > 0$;
   b. $\overline{A} > 0$, $\overline{X} > 0$, $\overline{B} < 0$;
   c. $\overline{A} > 0$, $\overline{X} < 0$, $\overline{B} > 0$;
   d. $\overline{A} > 0$, $\overline{X} < 0$, $\overline{B} < 0$;
   e. $\overline{A} < 0$, $\overline{X} > 0$, $\overline{B} < 0$;
   f. $\overline{A} < 0$, $\overline{X} < 0$, $\overline{B} > 0$;
   g. $\overline{A} < 0$, $\overline{X} < 0$, $\overline{B} < 0$.

4. In Example 5.3.1, find:

   a. $\overline{r}_e$;
   b. $\overline{r}_i$.
   d. And then compare them to $\overline{r}_c$ in Figure 5.3.

5. In Example 5.3.1 show that $\partial r_1/\partial\alpha > 0$, $\partial r_2/\partial\alpha < 0$ and $r_1(1) = r_2(1)$.

6. Let $\overline{A} = 1$ (crisp one), $\overline{B} = (2/4/7)$, $\overline{C} = 0$ (real zero) and $\overline{D} > 0$, in equation (5.26).

   a. Find $\overline{X}_c$, or show that it does not exist.
   b. Find $\overline{X}_e$.
   c. Find $\overline{X}_i$ and show $\overline{X}_e \leq \overline{X}_i$ ($\overline{X}_e(x) \leq \overline{X}_i(x)$, for all x).

7. Let $\overline{A} = 1$, $\overline{B} =$ zero, $\overline{C} = (0/0.5/1)$ and $\overline{D} = (0/4/8)$, in equation (5.26).

   a. Find $\overline{X}_c$, or show that it does not exist.
   b. Find $\overline{X}_e$.
   c. Show $\overline{X}_c \leq \overline{X}_e$ (if $\overline{X}_c$ exists).

8. Consider $\overline{A} \cdot \overline{X}^2 = \overline{D}$, $\overline{A} > 0$, and $\overline{D} > 0$. What conditions, if any, must $\overline{A}$ and $\overline{D}$ meet so that $\overline{X}_c$ exists. Find $\overline{X}_e$ and $\overline{X}_i$.

9. Consider $\overline{X}^2 + \overline{C} = \overline{D}$, $\overline{C} > 0$, $\overline{D} > 0$. Redo Problem 8 for these conditions on $\overline{C}$ and $\overline{D}$.

10. Consider the fuzzy quadratic when the classical solution does not exist and then look at the substitute classical solution $\overline{X}_s$ defined in Problem 10 in Section 5.2.4. Come up with an example of a fuzzy quadratic where $\overline{X}_c$ does not exist and then apply the method of Problem 10, Section 5.2.4, to solve for $\overline{X}_s$ and compare to $\overline{X}_e$ and $\overline{X}_i$.

Chapter 6

Fuzzy Inequalities

6.1 Introduction

In this chapter we start the solution to fuzzy inequalities. In the next section we consider solving fuzzy linear inequalities. The solution set depends on what "$\leq$ ", or " $<$ ", we will use between fuzzy numbers. In the text we use those described in Examples 4.5.1 to 4.5.3 in Chapter 4. We continue the development with discussing the solution set to fuzzy quadratic inequalities in the third section. Since the solution set to fuzzy inequalities is usually infinite, we try to describe this set for each problem and sometimes exhibit a particular member of the solution set. Applications of fuzzy inequalities would be in fuzzy optimization (fuzzy constraints) in Chapter 16.

6.2 Solving $\overline{A} \cdot \overline{X} + \overline{B} \leq \overline{C}$.

We wish to describe all solutions to

$$\overline{A} \cdot \overline{X} + \overline{B} \leq \overline{C} \ (or < \overline{C}), \tag{6.1}$$

for triangular fuzzy numbers $\overline{A}$, $\overline{B}$, $\overline{C}$. $\overline{X}$ will be a continuous fuzzy number.

In the crisp case

$$ax + b \leq c \tag{6.2}$$

we have $x \leq (c-b)/a$ for $a > 0$ and $x \geq (c-b)/a$ if $a < 0$. In either case there are an infinite number of solutions for x. The same will be true for equation (6.1). We need to be able to describe all continuous fuzzy numbers $\overline{X}$ which satisfy equation (6.1).

Let $\overline{E} = \overline{A} \cdot \overline{X} + \overline{B}$. First we need to decide how we are going to compute $\overline{E}$. We will use α-cuts and interval arithmetic for $\overline{E}$. Then the solution for $\overline{X}$ to $\overline{E} \leq \overline{C}$ ($\overline{E} < \overline{C}$) depends on your definition of "$\leq$" between fuzzy numbers. We will use the three definitions of "$\leq$" in Examples 4.5.1–4.5.3 in Section 4.5.

Example 6.2.1

Let $\overline{A} = (2/4/5)$, $\overline{B} = (-6/-3/-1)$ and $\overline{C} = (10/14/15)$. Since $\overline{A} > 0$ let us start with assuming $\overline{X} \approx (x_1/x_2, x_3/x_4)$, $\overline{X} > 0$. The α-cuts are $\overline{A}[\alpha] = [2+2\alpha, 5-\alpha]$, $\overline{B}[\alpha] = [-6+3\alpha, -1-2\alpha]$, $\overline{C}[\alpha] = [10+4\alpha, 15-\alpha]$, $\overline{X}[\alpha] = [x_1(\alpha), x_2(\alpha)]$. Then

$$\overline{E}[\alpha] = [(2+2\alpha)x_1(\alpha) + (-6+3\alpha), (5-\alpha)x_2(\alpha) + (-1-2\alpha)]. \tag{6.3}$$

Now we use "$\leq$" from Example 4.5.1 with $\theta = 0.8$. So we wish to solve $\overline{E} < \overline{C}$. Then $v(\overline{E} \leq \overline{C}) = 1$ and $v(\overline{C} \leq \overline{E}) < 0.8$ so that $\overline{E} < \overline{C}$. From Figure 6.1 we see that $\overline{X} \approx (x_1/x_2, x_3/x_4)$ is a solution if and only if: (1) $e_3 < 13.2$; and (2) the graph of $y = \overline{E}(x)$ on $[e_3, e_4]$ intersects $y = \overline{C}(x)$ at point Q below point P, or otherwise $e_4 \leq 10$. But $e_3 = 4x_3 - 3$ from equation (6.3) so $x_3 < 4.05$. $\overline{E}$ is trapezoidal shaped, $\overline{E} \approx (e_1/e_2, e_3/e_4)$ since $\overline{X}$ is trapezoidal shaped. Now we can put any curve in for $y = \overline{E}(x)$ on $[e_3, e_4]$ when $e_4 > 10$ as long as Q is below P, but also when we solve for $x_2(\alpha)$ we must have $dx_2/d\alpha < 0$.

For example if $\overline{E}[\alpha] = [e_1(\alpha), e_2(\alpha)]$ let $e_2(\alpha) = 14 - \alpha$. Then solving $14 - \alpha = 10 + 4\alpha$ we get 0.8. So $Q = P$. Try $e_2(\alpha) = 14 - 2\alpha$, then $14 - 2\alpha = 10 + 4\alpha$ gives $\alpha = 0.67$ and Q is below P. Now solve

$$(5-\alpha)x_2(\alpha) + (-1-2\alpha) = 14 - 2\alpha, \tag{6.4}$$

for $x_2(\alpha)$ giving

$$x_2(\alpha) = 15/(5-\alpha). \tag{6.5}$$

But $dx_2/d\alpha > 0$. Finally try $e_2(\alpha) = 14 - 6\alpha$. Then $14 - 6\alpha = 10 + 4\alpha$ gives $\alpha = 0.4$ and solving for $x_2(\alpha)$ produces

$$x_2(\alpha) = \frac{15 - 4\alpha}{5 - \alpha} \tag{6.6}$$

with $dx_2/d\alpha < 0$.

We summarize these results: $\overline{X} \approx (x_1/x_2, x_3/x_4)$, $\overline{X}[\alpha] = [x_1(\alpha), x_2(\alpha)]$ solves

$$\overline{A} \cdot \overline{X} + \overline{B} < \overline{C} \tag{6.7}$$

using "$<$" from Example 4.5.1 with $\overline{A} = (2/4/5)$, $\overline{B} = (-6/-3/-1)$, $\overline{C} = (10/14/15)$ if: (1) $x_3 < 4.05$; and (2) $x_3 < x_4 < 2.2$ ($e_4 \leq 10$) or $x_2(\alpha)$ is chosen so that $e_2(\alpha)$ intersects $y = \overline{C}(x)$ on $[e_3, e_4]$ below $\alpha = 0.8$ (See Figure 6.1).

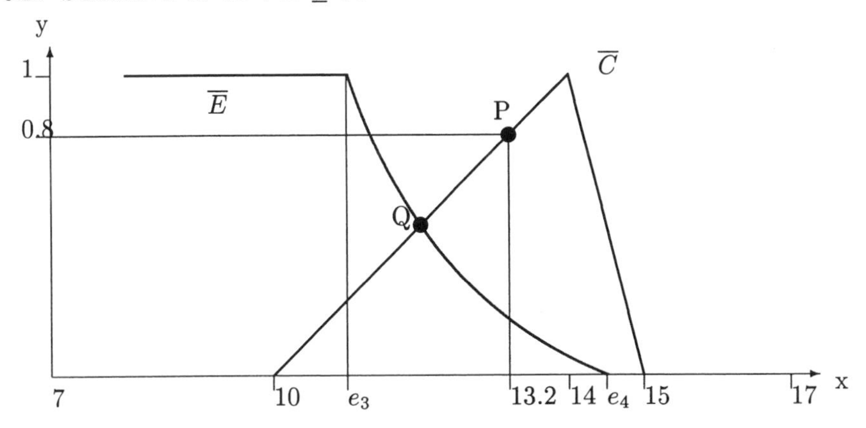

Figure 6.1: Solution to Example 6.2.1

Example 6.2.2

This continues Example 6.2.1 but now we will use the "$\leq$" from Example 4.5.2. We end up with $\overline{E} \leq \overline{C}$ as before but now this means $e_i(\alpha) \leq c_i(\alpha)$, $i = 1, 2$. In this problem we wish to solve $\overline{E} \leq \overline{C}$. From equation (6.3) we have

$$(2 + 2\alpha)x_1(\alpha) + (-6 + 3\alpha) \leq 10 + 4\alpha, \tag{6.8}$$

$$(5 - \alpha)x_2(\alpha) + (-1 - 2\alpha) \leq 15 - \alpha. \tag{6.9}$$

Or

$$x_1(\alpha) \leq \frac{16 + \alpha}{2 + 2\alpha} = l(\alpha), \tag{6.10}$$

$$x_2(\alpha) \leq \frac{16 + \alpha}{5 - \alpha} = r(\alpha). \tag{6.11}$$

The graphs of $l(\alpha)$ and $r(\alpha)$ are shown in Figure 6.2. Since $dx_1/d\alpha > 0$, $dx_2/d\alpha < 0$ and $x_1(1) \leq x_2(1)$ we conclude that the only restriction on $\overline{X} \approx (x_1/x_2, x_3/x_4)$, for it to be a solution is for $x_4 \leq 3.2$.

Example 6.2.3

This will continue Example 6.2.1 with "$\leq$" from Example 4.5.3. Solve $\overline{E} \leq \overline{C}$. We get $\overline{E} \leq \overline{C}$ if and only if the midpoint of the core of $\overline{E}$ is ≤ 14. The midpoint of the core of $\overline{E}$ is $2(x_2 + x_3) - 3$. Hence, $\overline{X}$ is a solution if and only if

$$x_2 + x_3 \leq 8.5 \tag{6.12}$$

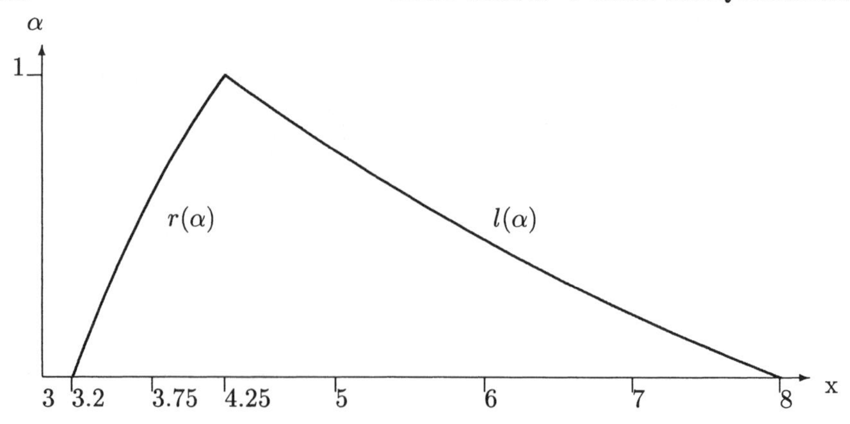

Figure 6.2: Constraints for the Solution to Example 6.2.2

6.3 $\overline{A} \cdot \overline{X}^2 + \overline{B} \cdot \overline{X} + \overline{C} \geq \overline{D}$ (or $> \overline{D}$).

Let $\overline{E} = \overline{A} \cdot \overline{X}^2 + \overline{B} \cdot \overline{X} + \overline{C}$ to be found using α-cuts and interval arithmetic. Then, as in the previous section, all we need to do is describe all $\overline{X} \approx (x_1/x_2, x_3/x_4)$ so that $\overline{E} \geq \overline{D}$ (or $> \overline{D}$).

Example 6.3.1

This example continues Example 6.2.1 where we use the ">" from Example 4.5.1. We wish to solve $\overline{E} > \overline{D}$. Assume that $\overline{X} > 0$. Let us use the same $\overline{A}$, $\overline{B}$, $\overline{C}$ as in Example 6.2.1 and set $\overline{D} = (-1/0/1)$. The α-cuts of $\overline{E}$ are:

$$e_1(\alpha) = (2 + 2\alpha)(x_1(\alpha))^2 + (-6 + 3\alpha)x_2(\alpha) + (10 + 4\alpha), \tag{6.13}$$

$$e_2(\alpha) = (5 - \alpha)(x_2(\alpha))^2 + (-1 - 2\alpha)x_1(\alpha) + (15 - \alpha). \tag{6.14}$$

Figure 6.3 shows the relationship between $\overline{E}$ and $\overline{D}$ for $\overline{E} > \overline{D}$. For $\overline{E} > \overline{D}$ we must have $0.2 < e_2$ and the graph of $e_1(\alpha)$ intersect $1 - \alpha$ below P whenever $e_1 < 1$. Now $0.2 < e_2$ means $0.2 < 4x_2^2 - 3x_3 + 14$ where $\overline{X} \approx (x_1/x_2, x_3/x_4)$. The other constraint is more difficult since it involves both $x_1(\alpha)$ and $x_2(\alpha)$ (equation (6.13)).

So, if $e_1 < 1$ and $0.2 < 4x_2^2 - 3x_3 + 14$ we choose $x_1(\alpha)$ and $x_2(\alpha)$ so that the solution to $e_1(\alpha) = 1 - \alpha$, call it α^*, has $\alpha^* < 0.8$. For example, let $x_1(\alpha) = 1 + 0.5\alpha$ and $x_2(\alpha) = 2 - 0.5\alpha$ so that $\overline{X} = (1/1.5/2)$. Then $e_1 = 0$, $4x_2^2 - 3x_3 + 14$ exceeds 0.2 and the intersection point Q in Figure 6.3 is less than 0.1. This $\overline{X}$ belongs to the solution set.

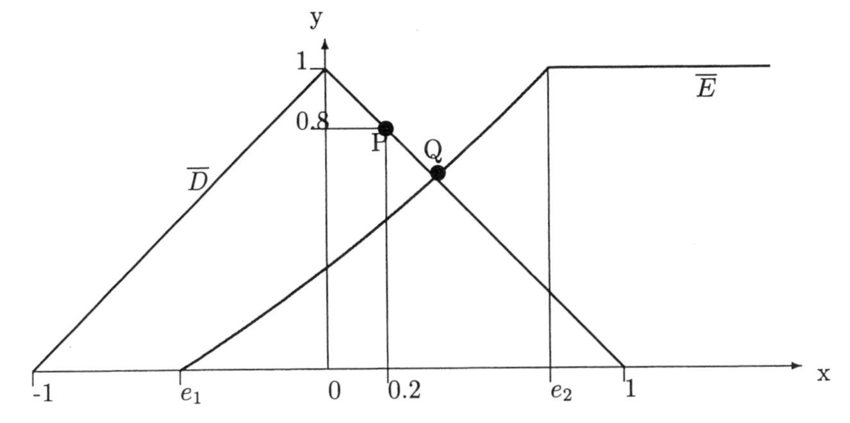

Figure 6.3: Solution for Example 6.3.1

Example 6.3.2

Same as Example 6.3.1 but use the "$\leq$" from Example 4.5.2. Now we want to solve $\overline{E} \geq \overline{D}$. So $\overline{X}$ is in the solution set when $e_1(\alpha) \geq -1 + \alpha$ and $e_2(\alpha) \geq 1 - \alpha$. Or

$$(2 + 2\alpha)(x_1(\alpha))^2 + (-6 + 3\alpha)x_2(\alpha) + (11 + 3\alpha) \geq 0, \tag{6.15}$$

$$(5 - \alpha)(x_2(\alpha))^2 + (-1 - 2\alpha)x_1(\alpha) + (14) \geq 0, \tag{6.16}$$

all $0 < \alpha \leq 1$. It is difficult to give general solutions for $x_1(\alpha)$ and $x_2(\alpha)$ from equations (6.15) and (6.16).

Example 6.3.3

This continues Example 6.3.1 but use the "$\leq$" from Example 4.5.3. Again we want to solve $\overline{E} > \overline{D}$. All we need is for the center of the core of $\overline{E}$ to be at least zero, the core of $\overline{D}$. That is, $\overline{X} \approx (x_1/x_2, x_3/x_4)$ is in the solution set if and only if

$$2(x_2^2 + x_3^2) - 1.5(x_2 + x_3) + 14 > 0. \tag{6.17}$$

6.3.1 Exercises

1. Work through the details of Example 6.2.1 if $\overline{X} < 0$.
2. Work through the details of Example 6.2.1 if $\overline{A} = (-5/-4/-2)$ and $\overline{X} < 0$.
3. Work through the details of Example 6.2.2 if $\overline{X} < 0$.
4. Work through the details of Example 6.2.2 if $\overline{A} = (-5/-4/-2)$ and $\overline{X} < 0$.
5. Work through the details of Example 6.2.3 if $\overline{X} < 0$.
6. Work through the details of Example 6.2.3 if $\overline{A} = (-5/-4/-2)$ and $\overline{X} < 0$.
7. Work through the details of Example 6.3.1 if $\overline{X} < 0$.
8. Work through the details of Example 6.3.1 if $\overline{A} = (-5/-4/-2)$ and $\overline{X} < 0$.
9. Work through Example 6.3.2 using $\overline{A} = (-5/-4/-2)$ and $\overline{X} < 0$.
10. Work through Example 6.3.3 using $\overline{A} = (-5/-4/-2)$ and $\overline{X} < 0$.
11. Discuss solving
$$\frac{\overline{X} + \overline{A}}{\overline{B}\overline{X} + 10} \leq \overline{C}$$
for $\overline{A} = (3/6/7)$, $\overline{B} = (14/20/26)$, $\overline{C} = (100/150/200)$, $\overline{X} > 0$ using "$\leq$" from Examples 4.5.1 - 4.5.3. Evaluate using α-cuts and interval arithmetic.
12. Discuss Example 6.2.1 using "$\leq$" from Example 4.5.4.
13. Discuss Example 6.3.1 using " $\leq$ " from Example 4.5.4.
14. Find three more $\overline{X}$ in the solution set in Example 6.2.1.
15. Find three more $\overline{X}$ in the solution set in Example 6.3.1.
16. Solutions to many simple crisp inequalities can be written using interval notation. For example, $-4 < 2x - 6 \leq 10$ can be written as $(1, 8]$ using interval notation. Can you define interval notation for continuous fuzzy numbers? Use " $\leq$ " from Examples 4.5.1 to 4.5.4 in Chapter 4. Then solve the examples in this chapter using this interval notation.

Chapter 7

Fuzzy Relations

7.1 Introduction

Fuzzy relations were briefly introduced in Section 3.2 and now we shall study them in more detail. Fuzzy relations are important in fuzzy systems theory and we shall use them again in Chapter 14. In the next section we present the basic definitions of crisp and fuzzy relations and concentrate on the definition and properties of the composition of fuzzy relations. In the third section we discuss reflexive, symmetric and transitive fuzzy relations and focus on the property of being transitive. If a fuzzy relation is not transitive we can form its transitive closure which requires finding powers of type I (all elements in $[0,1]$) fuzzy matrices. Therefore, we have to talk about sequences of powers of type I fuzzy matrices which are known to converge or oscillate. Section four is about fuzzy equivalence relations whose α-cuts give crisp equivalence relations. Solving fuzzy relational equations comprises the final section.

7.2 Definitions

We start with crisp (non-fuzzy) relations and then generalize to fuzzy relations. If X and Y are two sets, then $X \times Y$ is the set of all ordered pairs (x, y) for $x \in X$ and $y \in Y$. A crisp relation R between X and Y is a subset of $X \times Y$. So $R \subseteq X \times Y$. We use the notation from Chapter 2 for the characteristic function of R which means that $R(x, y) = 1$ if and only if $(x, y) \in R$. That is, $R(x, y) = 1$ if (x, y) is in R and $R(x, y) = 0$ for (x, y) not in R. $R(x, y) = 1$ means that x and y are related (associated) through relation R and $R(x, y) = 0$ means that they are not related. The inverse of R, written R^{-1}, is defined by $R^{-1}(x, y) = R(y, x)$.

Let R and S be two relations between X and Y. Since they are subsets of $X \times Y$ we may find $R \cup S$, $R \cap S$, R^c, etc.

Example 7.2.1

Let X be all people, aged 18 or more, in a certain town FUZZ. Let Y be all banks that have an office in FUZZ. Define $R_1(x,y) = 1$ if and only if x is a male in X who has an account with y in Y, and $R_2(x,y) = 1$ if and only if x is a female in X who has an account with $y \in Y$. If $S(x,y) = 1$ if and only if x is a senior (aged 55 or older) in X with an account with y in Y, then we might be interested in finding $R_1 \cup R_2$, $S \cap (R_1 \cup R_2)$, etc.

Example 7.2.2

Let $X = Y = \mathbf{R}$ so that R will be a subset of $\mathbf{R}^2$. Then R could be: (1) $R(x,y) = 1$ if and only if $x \leq y$; (2) $R(x,y) = 1$ if and only if $x^2 + y^2 \geq 1$; or (3) $R(x,y) = 1$ if and only if $xy \leq 0$.

A fuzzy relation $\overline{R}$ is just a fuzzy subset of $X \times Y$. So now $\overline{R}(x,y)$ can be any number in the interval $[0,1]$. $\overline{R}(x,y)$ gives the strength (from zero to one) of the relationship between x and y. The inverse of $\overline{R}$, written as $\overline{R}^{-1}$, is defined by $\overline{R}^{-1}(x,y) = \overline{R}(y,x)$. Since fuzzy relations $\overline{R}$, $\overline{S}$ and $\overline{T}$ are all fuzzy subsets of $X \times Y$, we can compute $\overline{R} \cup \overline{S}$, $\overline{R} \cap \overline{S}$, $\overline{R}^c$, $\overline{R} \cap (\overline{S} \cup \overline{T})$, etc. Also $\overline{R} \leq \overline{S}$ will mean that $\overline{R}(x,y) \leq \overline{S}(x,y)$ for all $(x,y) \in X \times Y$.

Example 7.2.3

If X and Y are finite sets, then $\overline{R}$ can be represented as a matrix or a diagram. Let $X = \{x_1, \cdots, x_m\}$ and $Y = \{y_1, \cdots, y_n\}$. Define $\overline{R}(x_i, y_j) = r_{ij} \in [0,1]$ for $1 \leq i \leq m$, $1 \leq j \leq n$. Then we may write $\overline{R}$ as a $m \times n$ matrix $[r_{ij}]$. For example

$$\overline{R} = \begin{pmatrix} 0.7 & 0.4 \\ 1.0 & 0.2 \\ 0.5 & 0.8 \end{pmatrix} \tag{7.1}$$

is a fuzzy relation between $X = \{x_1, x_2, x_3\}$ and $Y = \{y_1, y_2\}$ with $\overline{R}(x_1, y_2) = 0.4$, $\overline{R}(x_3, y_1) = 0.5$, etc. Figure 7.1 represents this $\overline{R}$ as a diagram.

If R is a crisp relation between X and Y, and S is a crisp relation between Y and Z, then the composition $R \circ S = T$ creates a new crisp relation between X and Z. The definition of T is $T(x,z) = 1$ if and only if there is a $y \in Y$ so that $R(x,y) = S(y,z) = 1$. We may write T as follows

$$T(x,z) = \max_y\{min(R(x,y), S(y,z))\}. \tag{7.2}$$

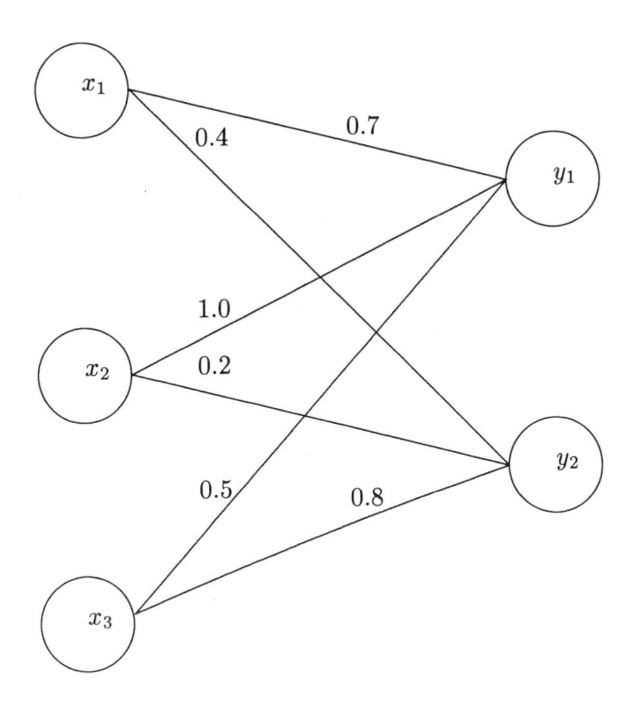

Figure 7.1: Fuzzy Relation as a Diagram

We used t-norm $T_m = \min$ in equation (7.2) but all t-norms could be used giving

$$T(x, z) = \max_y\{T(R(x, y), R(y, z))\}, \tag{7.3}$$

for any t-norm T. All t-norms will give the same result for $R \circ S$.

Now let $\overline{R}$ be a fuzzy relation on $X \times Y$ and $\overline{S}$ a fuzzy relation on $Y \times Z$. Then $\overline{T} = \overline{R} \circ \overline{S}$ is defined as

$$\overline{T}(x, z) = \sup_y\{T(\overline{R}(x, y), \overline{S}(y, z))\}, \tag{7.4}$$

giving $\overline{T}$ a fuzzy relation on $X \times Z$ for any t-norm T. The standard composition is when $T = T_m = \min$. We will usually use t-norm min in equation (7.4). All t-norms do not give the same result for fuzzy relations.

We would now like to determine the basic properties of composition of fuzzy relations. However, we must be careful because a certain property may be true for t-norm min but false for t-norm T_p. For example, using t-norm min in equation (7.4) we can show

$$(\overline{R} \circ \overline{S})^{-1} = \overline{S}^{-1} \circ \overline{R}^{-1}, \tag{7.5}$$

$$\overline{R} \circ (\overline{S} \circ \overline{T}) = (\overline{R} \circ \overline{S}) \circ \overline{T}, \tag{7.6}$$

$$\overline{R} \circ \overline{S} \neq \overline{S} \circ \overline{R}. \tag{7.7}$$

But these results may not hold for T_b, T_p, T^*. You are asked to investigate this further in the exercises.

When X, Y and Z are finite sets composition may be done in terms of matrices. Let X have m members, Y has n elements and Z contains k members. Also let $\overline{R} = [r_{ij}]$ be a $m \times n$ matrix for a fuzzy relation between X and Y, $\overline{S} = [s_{jl}]$ a $n \times k$ matrix for the fuzzy relation between Y and Z. If $\overline{T} = \overline{R} \circ \overline{S}$, then $\overline{T} = [t_{il}]$ a $m \times k$ matrix for the new fuzzy relation between X and Z. Using t-norm min we may find the t_{il} as

$$t_{il} = \max\{min(r_{i1}, s_{1l}), \cdots, min(r_{in}, s_{nl})\}. \tag{7.8}$$

Usual matrix multiplication is

$$[t_{il}] = [r_{ij}][s_{jl}], \tag{7.9}$$

$$t_{il} = \sum_{j=1}^{n} r_{ij} s_{jl}. \tag{7.10}$$

But in fuzzy relation composition using t-norm min, you replace multiplication with min and addition with max. If we used T_p, we only replace addition with max.

Example 7.2.4

Let

$$\overline{R} = \begin{pmatrix} 0.7 & 0.4 \\ 1.0 & 0.2 \\ 0.5 & 0.8 \end{pmatrix} \tag{7.11}$$

and

$$\overline{S} = \begin{pmatrix} 0.4 & 1.0 \\ 1.0 & 0.3 \end{pmatrix} \tag{7.12}$$

Then if $\overline{T} = \overline{R} \circ \overline{S}$, using t-norm T_m, we calculate

$$\overline{T} = \begin{pmatrix} 0.4 & 0.7 \\ 0.4 & 1.0 \\ 0.8 & 0.5 \end{pmatrix} \tag{7.13}$$

For example

$$t_{21} = 0.4 = \max\{\min(1.0, 0.4), \min(0.2, 1.0)\}. \tag{7.14}$$

7.2.1 Exercises

1. If X and Y are finite sets and X has m elements, Y has n members, then how many distinct crisp relations R are there between X and Y?

2. Argue that all t-norms, in equation (7.3), will give the same result for relation T.

3. Determine if the following equation is true or false for the given t-norm and t-conorm. To show an equation is not true we suggest you consider finite sets for X,Y, Z and W. $\overline{R}$ is a fuzzy relation between X and Y, $\overline{S}$ is between Y and Z, and $\overline{T}$ is between Z and W.

   a. $(\overline{R} \circ \overline{S})^{-1} = \overline{S}^{-1} \circ \overline{R}^{-1}$.
   b. $(\overline{R} \circ \overline{S}) \circ \overline{T} = \overline{R} \circ (\overline{S} \circ \overline{T})$.
   c. $\overline{R} \circ \overline{S} = \overline{S} \circ \overline{R}$.
   d. $\overline{R} \circ (\overline{S}_1 \cup \overline{S}_2) = \overline{R} \circ \overline{S}_1 \cup \overline{R} \circ \overline{S}_2$.
   e. $\overline{R} \circ (\overline{S}_1 \cap \overline{S}_2) = \overline{R} \circ \overline{S}_1 \cap \overline{R} \circ \overline{S}_2$.
   f. $(\overline{R}_1 \cup \overline{R}_2) \circ \overline{S} = \overline{R}_1 \circ \overline{S} \cup \overline{R}_2 \circ \overline{S}$.
   g. $(\overline{R}_1 \cap \overline{R}_2) \circ \overline{S} = \overline{R}_1 \circ \overline{S} \cap \overline{R}_2 \circ \overline{S}$.
   h. $\overline{R}_1 \leq \overline{R}_2$ implies $\overline{R}_1 \circ \overline{S} \leq \overline{R}_2 \circ \overline{S}$.
   i. $\overline{S}_1 \leq \overline{S}_2$ implies $\overline{R} \circ \overline{S}_1 \leq \overline{R} \circ \overline{S}_2$.
   j. $\overline{R}^c \cup \overline{R} = \overline{X}^2$ where $\overline{X}^2(x,y) = 1$ all x, y .
   k. $\overline{R}^c \cap \overline{R} = \overline{\phi}$ where $\overline{\phi}(x,y) = 0$ all x, y.
   l. If $X = Y = Z = W$ and $\cdot$ is multiplication, then $\overline{R} \cdot (\overline{S} \cup \overline{T}) = (\overline{R} \cdot \overline{S}) \cup (\overline{R} \cdot \overline{T})$.
   m. If $X = Y = Z$, then $\overline{R} \circ \overline{S}^c = \overline{R} - \overline{R} \circ \overline{S}$.

   Use the following t-norms and their dual t-conorms. If you use t-norm T for composition, then use the same t-norm (or its dual t-conorm) for intersection (union) of fuzzy sets.

   i. t-norm T_m and t-conorm C_m.
   ii. t-norm T_b and t-conorm C_b.
   iii. t-norm T_p and t-conorm C_p.
   iv. t-norm T^* and t-conorm C^*.

4. Let

$$\overline{R}(x,y) = \begin{cases} \sqrt{4-(x-1)^2-(y-1)^2}, & (x-1)^2+(y-1)^2 \leq 4 \\ 0, & \text{otherwise.} \end{cases}$$

   $\overline{R}$ represents the idea of "near" the point (1,1) in $\mathbf{R}^2$. Evaluate :

a. $\overline{R}(0,0)$;
b. $\overline{R}(1,0)$;
c. $\overline{R}(0,1)$;
d. $\overline{R}(1,1)$.

5. In Example 7.2.1 describe in words the relations $\overline{R}_1 \cup \overline{R}_2$, $\overline{S} \cap (\overline{R}_1 \cup \overline{R}_2)$ and $\overline{S}^c \cap (\overline{R}_1 \cup \overline{R}_2)$.

6. Compute $\overline{T} = \overline{R} \circ \overline{S}$ in Example 7.2.4 using t-norm:

   a. T_b;
   b. T_p;
   c. T^*.

7. Let

$$\overline{R} = \begin{pmatrix} 0.2 & 0 & 0.5 \\ 0.3 & 0.2 & 1 \\ 0 & 0.5 & 0.6 \end{pmatrix}$$

and

$$\overline{S} = \begin{pmatrix} 0.3 & 0.6 & 0 \\ 0.9 & 1 & 0.4 \\ 1 & 0.6 & 0.5 \end{pmatrix}$$

Find the max-min and the max-product ($T = T_p$) composition of $\overline{R}$ and $\overline{S}$.

8. Let R be a crisp relation on $X \times Y$. Define

$$R_x = \{x \in X | (x,y) \in R \textit{ for some } y \in Y\},$$

$$R_y = \{y \in Y | (x,y) \in R \textit{ for some } x \in X\}.$$

Show that:

   a. the characteristic function of $R_x = \sup_y R(x,y)$;
   b. the characteristic function of $R_y = \sup_x R(x,y)$;
   c. $R \subseteq R_x \times R_y$.

9. Define level 2 fuzzy relations (see Section 3.2).

   a. Give some "real world" applications of this type of fuzzy relation.
   b. Define composition between these types of fuzzy relations.

10. Define type 2 fuzzy relations (see Section 3.2).

   a. Give some "real world" applications of this type of fuzzy relation.
   b. Define composition between these types of fuzzy relations.

7.3 Transitive Closure

In this section $X = Y$ and X is a finite set. Let $X = \{x_1, \cdots, x_n\}$ and R be a crisp relation on X. When $X = Y$, we will say R is a relation on X instead of saying that R is a relation between X and X $(Y = X)$. We may also use the terminology of R is a relation on $X \times X$ when $X = Y$. Basic properties a relation R may, or may not possess, are:

1. reflexive : $R(x, x) = 1$, all $x \in X$;
2. symmetric: $R(x, y) = R(y, x)$, all $x, y \in X$;
3. transitive: $R(x, y) = R(y, z) = 1$ implies $R(x, z) = 1$.

Example 7.3.1

Assume $X = \{x_1, x_2, x_3, x_4\}$ and consider the R given by the matrix

$$R = \begin{pmatrix} 1 & 1 & 0 & 1 \\ 1 & 1 & 0 & 1 \\ 0 & 0 & 1 & 1 \\ 1 & 1 & 1 & 1 \end{pmatrix} \tag{7.15}$$

where $R(x_1, x_4) = 1$, $R(x_3, x_2) = 0$, etc. R is reflexive since $R(x_i, x_i) = 1$, $1 \leq i \leq 4$. Notice also that R is symmetric because $R(x_1, x_2) = R(x_2, x_1), \cdots, R(x_3, x_4) = R(x_4, x_3)$. Now check to see if R is transitive. There are many things to check but we finally come up with $R(x_3, x_4) = R(x_4, x_2) = 1$ but $R(x_3, x_2) = 0$. R is not transitive.

Another way to check to see if R is transitive is to compute $R \circ R$ and then R is transitive when

$$R = R \circ R. \tag{7.16}$$

If R is not transitive one can find its transitive closure R_T. The transitive closure of R is defined as follows:

1. $R \subset R_T$;
2. R_T is transitive ;
3. if $R \subset S$ and S is transitive, then $R_T \subseteq S$.

Therefore R_T is the "smallest" transitive relation containing R. We will not study the algorithm used to find R_T and instead we will do this for fuzzy relations.

The basic properties of a fuzzy relation $\overline{R}$ are :

1. reflexive: $\overline{R}(x,x) = 1$ all $x \in X$;

2. symmetric: $\overline{R}(x,y) = \overline{R}(y,x)$ all $x, y \in X$;

3. transitive: $\overline{R} \circ \overline{R} \leq \overline{R}$.

Let us rewrite the transitive equation for fuzzy relation

$$\max_{y}\{min(\overline{R}(x,y), \overline{R}(y,z))\} \leq \overline{R}(x,z). \tag{7.17}$$

We used t-norm min in equation (7.17) but we could also use any other t-norm. In the future we always need to point out which t-norm we are using in the definition of transitive for fuzzy relations. For crisp relations it is $R \circ R = R$ for transitive but for fuzzy relations we require $\overline{R} \circ \overline{R} \leq \overline{R}$.

$\overline{R} \geq \overline{R} \circ \overline{R}$ means that the strength of the link between x and z is not less than the maximum strength of paths from x to z through any other point y. Choose $x, z \in X$ and look at all paths from x to z through an intermediary point y. So we go from x to y to z and the strength of this path is $min(\overline{R}(x,y), \overline{R}(y,z))$, the strength of its minimum connection. Now $\overline{R} \circ \overline{R}$ finds the maximum of the strengths of all paths x to y to z. We do it this way for fuzzy relations because $\overline{R}$ values can be any number in [0,1] whereas R values are only zero or one.

Now we want to discuss the transitive closure of a fuzzy relation assuming, of course, that it is not transitive to start with. But first we need to study powers of type I fuzzy matrices. A type I fuzzy matrix has all its elements numbers in the interval [0,1]. A type II fuzzy matrix has all its elements fuzzy numbers. Type I fuzzy matrices arise from fuzzy relations on finite sets. We will use type II fuzzy matrices in Chapter 11. Let $\overline{R}(x_i, x_j) = r_{ij}$ for $1 \leq i, j \leq n$, for fuzzy relation $\overline{R}$ on X and X has n members. Then we write $\overline{R} = [r_{ij}]$, a $n \times n$ matrix whose elements $r_{ij} \in [0,1]$. Now define the sequence $\overline{R}^2 = \overline{R} \circ \overline{R}$, $\overline{R}^3 = \overline{R}^2 \circ \overline{R}$, $\cdots$, $\overline{R}^{n+1} = \overline{R}^n \circ \overline{R}$, $\cdots$ using t-norm min.

It is known that the sequence $\overline{R}, \overline{R}^2, \cdots, \overline{R}^n, \cdots$ either converges or oscillates. By converge we mean that there is a positive integer p so that $\overline{R}^n = \overline{R}^p$ for $n \geq p$. This means that the sequence becomes $\overline{R}, \overline{R}^2, \cdots, \overline{R}^p, \overline{R}^p, \overline{R}^p, \cdots$ where $\overline{R}, \overline{R}^2, \cdots, \overline{R}^p$ are all different but past the first $\overline{R}^p$ they are all the same and equal to $\overline{R}^p$. By oscillate we mean there are $n \times n$ fuzzy matrices $\overline{S}_1, \cdots, \overline{S}_L$ and a positive integer p so that $\overline{R}^p = \overline{S}_1, \overline{R}^{p+1} = \overline{S}_2, \cdots, \overline{R}^{p+L-1} = \overline{S}_L, \overline{R}^{p+L} = \overline{S}_1, \cdots$. The sequence $\overline{R}, \overline{R}^2, \cdots, \overline{R}^p$ are all different but then the sequence is $\overline{S}_2, \cdots, \overline{S}_L, \overline{S}_1, \cdots, \overline{S}_L, \cdots$ forever. These results hold for t-norm $T_m = min$ in computing $\overline{R}^2, \overline{R}^3, \cdots$. You are asked to find out what happens to this sequence if you use another t-norm in the exercises.

Example 7.3.2

Let

$$\overline{R} = \begin{pmatrix} 1 & 0.2 & 0.8 \\ 0.5 & 1 & 0 \\ 0.7 & 0.6 & 1 \end{pmatrix} \tag{7.18}$$

Using t-norm min to find $\overline{R}^n$, $n \geq 2$, we obtain

$$\overline{R}^n = \begin{pmatrix} 1 & 0.6 & 0.8 \\ 0.5 & 1 & 0.5 \\ 0.7 & 0.6 & 1 \end{pmatrix} \tag{7.19}$$

for $n \geq 2$. The sequence $\overline{R}, \overline{R}^2, \cdots$ converges to the fuzzy matrix in equation (7.19).

Example 7.3.3

Let

$$\overline{R} = \begin{pmatrix} 0 & 0.2 & 1 \\ 0.4 & 0 & 1 \\ 0 & 1 & 0.3 \end{pmatrix} \tag{7.20}$$

We find, using t-norm min for the composition of fuzzy matrices, that the sequence $\overline{R}, \overline{R}^2, \cdots$ becomes $\overline{R}, \overline{R}^2, \cdots, \overline{R}^5, \overline{R}^6, \overline{R}^5, \overline{R}^6, \cdots$. That is, the sequence oscillates between $\overline{R}^5$ and $\overline{R}^6$ after $\overline{R}, \overline{R}^2, \overline{R}^3, \overline{R}^4$.

Now we may discuss the transitive closure of a fuzzy relation $\overline{R}$ on X. The properties of the transitive closure $\overline{R}_T$ are: (1) $\overline{R}_T$ is transitive; (2) $\overline{R} \leq \overline{R}_T$; and (3) if $\overline{S}$ is transitive with $\overline{R} \leq \overline{S}$, then $\overline{R}_T \leq \overline{S}$. In this sense $\overline{R}_T$ is the smallest transitive fuzzy relation containing $\overline{R}$. The formula for $\overline{R}_T$ is

$$\overline{R}_T = \overline{R} \cup \overline{R}^2 \cup \overline{R}^3 \cup \cdots \tag{7.21}$$

In equation (7.21) we must decide on the t-conorm to use for union and the t-norm to use for $\overline{R} \circ \overline{R}$. Let us use t-conorm $C_m = \max$ for union and t-norm min for composition. Then we can argue that the sequence in equation (7.21) must terminate. That is, there is a positive integer p so that

$$\overline{R}_T = \overline{R} \cup \overline{R}^2 \cup \cdots \cup \overline{R}^p. \tag{7.22}$$

If the sequence $\overline{R}, \overline{R}^2, \cdots$ converges, then $\overline{R}^n = \overline{R}^p$ for $n \geq p$, and we get equation (7.22) because we are using max for union. So suppose the sequence oscillates between $\overline{S}_1, \overline{S}_2$ and $\overline{S}_3$ starting at $\overline{R}^{p-2}$. Then equation (7.22) becomes

$$\overline{R}_T = \overline{R} \cup \cdots \cup \overline{R}^{p-3} \cup \overline{S}_1 \cup \overline{S}_2 \cup \overline{S}_3, \tag{7.23}$$

since we use max for union. Therefore , we obtain equation (7.22) for $\overline{R}^p = \overline{S}_3$.

Is $\overline{R}_T$ in equation (7.22) transitive? Using equation (7.22) we may show that $\overline{R}_T \circ \overline{R}_T \leq \overline{R}_T$ which implies it is transitive. Also, we can argue that given transitive $\overline{S}$, $\overline{R} \leq \overline{S}$, then $\overline{R}_T \leq \overline{S}$.

Example 7.3.4

Using the $\overline{R}$ from Example 7.3.2 we find $\overline{R}_T = \overline{R} \cup \overline{R}^2$, or

$$\overline{R}_T = \begin{pmatrix} 1 & 0.6 & 0.8 \\ 0.5 & 1 & 0.5 \\ 0.7 & 0.6 & 1 \end{pmatrix} \tag{7.24}$$

$\overline{R}_T$ is transitive because we see that $\overline{R}_T \circ \overline{R}_T = \overline{R}_T$.

7.3.1 Exercises

1. Let R be a crisp relation on $X \times X$. Show that :

   a. if $R \circ R = R$, then R is transitive ;

   b. if R is transitive and reflexive , then $R \circ R = R$.

2. Let $\overline{R}$ be a fuzzy relation on $X \times X$. Show that if $\overline{R}$ is transitive using t-norm T_m, then it is also transitive using t-norm T_p.

3. Show that $\overline{R}^{-1}$ is transitive if $\overline{R}$ is transitive.

4. Investigate the truth of all the statements " If $\overline{R}$ is transitive using t-norm T_α, then $\overline{R}$ is transitive using t-norm T_β " for $T_\alpha, T_\beta \in \{T_m, T_b, T_p, T^*\}$.

5. Determine what happens to the sequence $\overline{R}, \overline{R}^2, \cdots$ for $\overline{R}$ in Example 7.3.2 if we use the following t-norm for computing $\overline{R} \circ \overline{R}$.

   a. T_b.

   b. T_p.

   c. T^*.

6. Determine what happens to the sequence $\overline{R}, \overline{R}^2, \cdots$ for $\overline{R}$ in Example 7.3.3 if we use the following t-norms for finding $\overline{R} \circ \overline{R}$:

   a. T_b,

   b. T_p,

   c. T^*.

7. Show that $\overline{R}_T \circ \overline{R}_T \leq \overline{R}_T$ for $\overline{R}_T$ given in equation (7.22).

8. For $\overline{R}_T$ given in equation (7.22) show that $\overline{R}_T \leq \overline{S}$ for any transitive $\overline{S}$, $\overline{R} \leq \overline{S}$.

9. Determine what happens to the sequence in equation (7.21), finding the transitive closure $\overline{R}_T$ of fuzzy relation $\overline{R}$, if we use t-conorm C for union and t-norm T for composition, where:

   a. $C = C_b, T = T_b$;

   b. $C = C_p, T = T_p$;

   c. $C = C^*, T = T^*$.

10. Suppose $\overline{R}$ is reflexive. Show that the sequence $\overline{R}, \overline{R}^2, \cdots$ must converge, it can not oscillate, using t-norm T_m for composition. Is the same for t-norm T_p in computing $\overline{R} \circ \overline{R}$?

11. Find $\overline{R}_T$ in Example 7.3.2 using t-conorm C for union and t-norm T for composition if:

    a. $C = C_b, T = T_b$;
    b. $C = C_p, T = T_p$;
    c. $C = C^*, T = T^*$.

12. Find $\overline{R}_T$ for the $\overline{R}$ in Example 7.3.3 using t-conorm C in union and t-norm T for composition if:

    a. $C = C_m, T = T_m$;
    b. $C = C_p, T = T_p$.

13. Let $\overline{R}$ be reflexive. By Problem 10 the sequence $\overline{R}, \overline{R}^2, \cdots$ converges to $\overline{R}^p$, for some positive integer p. Show that $\overline{R}_T = \overline{R}^p$.

14. Show that if $\overline{R}$ is symmetric, then $\overline{R}^{-1} = \overline{R}$.

15. If $\overline{R}$ and $\overline{S}$ are symmetric fuzzy relations on the same set, is $\overline{R} \cap \overline{S}$ also symmetric, using t-norm min for intersection? If we use t-conorm max for union, will $\overline{R} \cup \overline{S}$ be symmetric?

16. If $\overline{R}$ is transitive using t-norm T_m, then show that all its α-cuts $\overline{R}[\alpha]$ define crisp transitive relations. Is this true if we use t-norm T_b, T_p, T^*? If $\overline{R}$ is symmetric (reflexive), then are its α-cuts also symmetric (reflexive)?

17. If $\overline{R}$ and $\overline{S}$ are two fuzzy relations on the same set X and $\overline{R}$ is reflexive, then show that $\overline{R} \cup \overline{S}$, using t-conorm max for union, is also reflexive.

18. Let R be a crisp relation on $X \times X$, with X a finite set. Assume R is not transitive and R_T is the transitive closure of R. An algorithm for R_T is

    a. set $R_1 = R \cup (R \circ R)$
    b. Does $R = R_1$?
        i. if yes, the $R_T = R_1$
        ii. if no, set $R = R_1$, go to a.

    Use t-norm T_m for $R \circ R$ and regular union for crisp sets. Show that R_T is transitive and $R \subset R_T$. Find R_T for the R in Example 7.3.1.

19 Let $\overline{R}$ and $\overline{S}$ be transitive fuzzy relations on X. Use T = min for intersection and C = max for union.

    a. Is $\overline{R} \cup \overline{S}$ transitive?
    b. Is $\overline{R} \cap \overline{S}$ transitive?

20. Define crisp relation R on the set of fuzzy numbers as $R(\overline{M}, \overline{N}) = 1$ if and only if $\overline{M}(x) \leq \overline{N}(x)$ for all x. Is R transitive?

7.4 Fuzzy Equivalence Relation

We start with a crisp relation R on $X \times X$. R is said to be an equivalence relation if it is reflexive, symmetric and transitive. An important property of equivalence relations is that they may be used to decompose X into a set of disjoint equivalence classes. We did this in propositional logic in Section 2.3.

For any $x \in X$ define

$$[x] = \{y|R(x,y) = 1\}, \tag{7.25}$$

and this crisp set $[x]$ is the equivalence class generated by x. Let

$$X/R = \{[x]|x \in X\}, \tag{7.26}$$

be the set of equivalence classes generated by R. We can show that:

1. any $x \in X$ belongs to one, and only one, equivalence class in X/R;
2. if a and b belong to $[x]$, then $R(a,b) = 1$, or a and b are related; and
3. if $[x] \neq [y]$, then $[x] \cap [y] = \phi$ and if $a \in [x]$, $b \in [y]$, then $R(a,b) = 0$.

It is easy to show #1 above since x belongs to $[x]$ because R is reflexive. The rest of #1 follows from #3, after we prove #3. To show #2 we have $R(x,a) = R(x,b) = 1$, or $R(a,x) = R(b,x) = 1$ by symmetry and by transitivity $R(a,b) = 1$. If $[x] \neq [y]$, then assume that $a \in [x] \cap [y]$. Then by symmetry and transitivity $R(x,y) = 1$ and it follows that every element in $[x]$ is related every member of $[y]$ and $[x] = [y]$, a contradiction. Hence $[x] \cap [y] = \phi$.

Example 7.4.1

Let

$$R = \begin{pmatrix} 1 & 0 & 1 & 0 \\ 0 & 1 & 0 & 1 \\ 1 & 0 & 1 & 0 \\ 0 & 1 & 0 & 1 \end{pmatrix} \tag{7.27}$$

which is an equivalence relation on finite set $X = \{x_1, x_2, x_3, x_4\}$.

We find

$$[x_1] = \{x_1, x_3\}, \tag{7.28}$$

and

$$[x_2] = \{x_2, x_4\}. \tag{7.29}$$

A fuzzy relation $\overline{R}$ on $X \times X$ which is reflexive, symmetric and transitive is called a fuzzy equivalence relation. Let us assume we are using t-norm

T_m to calculate $\overline{R} \circ \overline{R}$. Using t-norm min we have $\overline{R} \circ \overline{R} \leq \overline{R}$ for transitive $\overline{R}$. We now show that the α-cuts of $\overline{R}$ give crisp equivalence relations and equivalence classes. The α-cuts are

$$\overline{R}[\alpha] = \{(x,y) \in X \times X | \overline{R}(x,y) \geq \alpha\}, \tag{7.30}$$

for $0 < \alpha \leq 1$. Define crisp R_α on $X \times X$ as $R_\alpha(x,y) = 1$ if and only if $(x,y) \in \overline{R}[\alpha]$. We can argue that R_α is a crisp equivalence relation on $X \times X$ for all $0 < \alpha \leq 1$. Then we find X/R_α and look at the relationship between X/R_{α_1} and X/R_{α_2} for $0 < \alpha_1 < \alpha_2 \leq 1$.

Let us show why R_α is an equivalence relation on $X \times X$. First, $R_\alpha(x,x) = 1$, and R_α is reflexive, since $(x,x) \in \overline{R}[\alpha]$ all α because $\overline{R}(x,x) = 1$. We also have $R_\alpha(x,y) = R_\alpha(y,x)$ because: (1) if $(x,y) \in \overline{R}[\alpha]$, so is $(y,x) \in \overline{R}[\alpha]$ since $\overline{R}$ is symmetric and $R_\alpha(x,y) = R_\alpha(y,x) = 1$; and (2) if (x,y) is not in $\overline{R}[\alpha]$, then neither is (y,x) in $\overline{R}[\alpha]$ because $\overline{R}$ is symmetric, so that $R_\alpha(x,y) = R_\alpha(y,x) = 0$. Lastly, R_α is transitive because $\overline{R} \circ \overline{R} \leq \overline{R}$. Let $R_\alpha(x,y) = R_\alpha(y,z) = 1$. This means that $\overline{R}(x,y) \geq \alpha$ and $\overline{R}(y,z) \geq \alpha$. So $\min(\overline{R}(x,y), \overline{R}(y,z)) \geq \alpha$. Therefore, the supremum of $\min(\overline{R}(x,y), \overline{R}(y,z))$ over all $y \in X$ also is at least α. This means, by the transitivity of $\overline{R}$, that $\overline{R}(x,z) \geq \alpha$ so that $R_\alpha(x,z) = 1$ and R_α is transitive.

Next we wish to investigate the relationships between the equivalence classes, or X/R_α, as α increases from zero to one.

Example 7.4.2

Let

$$\overline{R} = \begin{pmatrix} 1 & 0.2 & 1 & 0.6 & 0.2 \\ 0.2 & 1 & 0.2 & 0.2 & 0.8 \\ 1 & 0.2 & 1 & 0.6 & 0.2 \\ 0.6 & 0.2 & 0.6 & 1 & 0.2 \\ 0.2 & 0.8 & 0.2 & 0.2 & 1 \end{pmatrix} \tag{7.31}$$

be a fuzzy relation on finite set X; X has 5 members. We check and see that $\overline{R}$ is reflexive, symmetric and transitive ($\overline{R} \circ \overline{R} = \overline{R}$). We obtain different R_α for α in the intervals $(0, 0.2], (0.2, 0.6], (0.6, 0.8], (0.8, 1]$. For α in $(0, 0.2]$ we obtain $R_\alpha = [r_{ij}]$ with $r_{ij} = 1$ for all i, j. The equivalence classes for this R_α is just one set X. So let α be in $(0.2, 0.6]$. Then

$$R_\alpha = \begin{pmatrix} 1 & 0 & 1 & 1 & 0 \\ 0 & 1 & 0 & 0 & 1 \\ 1 & 0 & 1 & 1 & 0 \\ 1 & 0 & 1 & 1 & 0 \\ 0 & 1 & 0 & 0 & 1 \end{pmatrix} \tag{7.32}$$

with equivalence classes

$$[x_1] = \{x_1, x_3, x_4\}, \tag{7.33}$$

and

$$[x_2] = \{x_2, x_5\}. \tag{7.34}$$

Next let α belong to $(0.6, 0.8]$ and we compute

$$R_\alpha = \begin{pmatrix} 1 & 0 & 1 & 0 & 0 \\ 0 & 1 & 0 & 0 & 1 \\ 1 & 0 & 1 & 0 & 0 \\ 0 & 0 & 0 & 1 & 0 \\ 0 & 1 & 0 & 0 & 1 \end{pmatrix} \tag{7.35}$$

with equivalence classes

$$[x_1] = \{x_1, x_3\}, \tag{7.36}$$

$$[x_2] = \{x_2, x_5\}, \tag{7.37}$$

and

$$[x_4] = \{x_4\}. \tag{7.38}$$

Finally, if α is in $(0.8, 1]$ we find R_α with equivalence classes

$$[x_1] = \{x_1, x_3\}, \tag{7.39}$$

$$[x_2] = \{x_2\}, \tag{7.40}$$

$$[x_4] = \{x_4\}, \tag{7.41}$$

and

$$[x_5] = \{x_5\}. \tag{7.42}$$

We notice that the equivalence classes tend to split up, and become smaller, as we increase α from zero to one.

7.4.1 Exercises

In Problems 7, 9–16 and 20 we use the notation $\mathcal{N}_0$ for the set of continuous fuzzy numbers and $\Upsilon = \mathcal{N}_0 \times \mathcal{N}_0$.

1. Let R be an equivalence relation on $X \times X$. If $[x] \neq [y]$ and $a \in [x]$, $b \in [y]$, then show that $R(a,b) = 0$.

2. Let $\overline{R}$ be a fuzzy equivalence relation on $X \times X$ using t-norm T_β for $\overline{R} \circ \overline{R}$. Define R_α on $\overline{R}[\alpha]$ as in this section. Is R_α a crisp equivalence relation on $X \times X$ for:

   a. $T_\beta = T_b$,

   b. $T_\beta = T_p$,

   c. $T_\beta = T^*$.

3. Let $\overline{R}$ be a fuzzy equivalence relation on $X \times X$ using t-norm T_m for $\overline{R} \circ \overline{R}$. Describe in detail the relationship between the equivalence classes generated by R_α and R_β if $0 < \alpha < \beta \leq 1$.

4. Is the $\overline{R}$ given in Example 7.4.2 also transitive if we use t-norm T_b, T_p or T^* for composition? If so, then find X/R_α as α increases from zero to one.

5. Let A_i, $1 \leq i \leq n$ be a crisp, non-empty, partition of set X. That is, $A_i \cap A_j = \phi$ for $i \neq j$ and $\bigcup A_i = X$. Define a crisp relation R on $X \times X$ as follows: (1) $R(x,y) = 1$ if x and y belong to the same A_i; and (2) $R(x,y) = 0$ if x and y belong to different A_i. Show that R is an equivalence relation on $X \times X$.

6. Let $X = \mathbf{R}$ and define a fuzzy relation $\overline{R}$ on $X \times X$ as follows: $R(x,y) = \exp(-|x-y|)$. Use t-norm T_p for composition. Is $\overline{R}$ a fuzzy equivalence relation on $X \times X$?

7. Let $\mathcal{N}_0$ be the set of all continuous fuzzy numbers and let ψ be any function mapping $\mathcal{N}_0$ into the real numbers. For example, $\psi(\overline{A})$ could be the midpoint of the core of $\overline{A}$. Define a crisp relation on $\mathcal{N}_0 \times \mathcal{N}_0$ as

$$R(\overline{A}, \overline{B}) = \begin{cases} 1, & \psi(\overline{A}) = \psi(\overline{B}), \\ 0, & \text{otherwise.} \end{cases}$$

   Is R an equivalence relation?

8. Let $\overline{R}$ and $\overline{S}$ be fuzzy equivalence relations on $X \times X$. Determine if the following are also fuzzy equivalence relations. Use $T = \min$ for composition and intersection and $C = \max$ for union.

   a. $\overline{R} \cap \overline{S}$.

b. $\overline{R} \circ \overline{S}$.

c. $\overline{R} \cup \overline{S}$.

9. Let $\Upsilon = \mathcal{N}_0 \times \mathcal{N}_0$. Define R on $\Upsilon \times \Upsilon$ as below. Determine if R is an equivalence relation. If so, describe the equivalence classes.

   a. $R((\overline{A}, \overline{B}), (\overline{M}, \overline{N})) = 1$ if and only if $\overline{A} \cdot \overline{M} = \overline{B} \cdot \overline{N}$.

   b. $R((\overline{A}, \overline{B}), (\overline{M}, \overline{N})) = 1$ if and only if $\overline{A} + \overline{M} = \overline{B} + \overline{N}$.

10. Let $\overline{O} = (-0.1/0/0.1)$. Also let $D(\overline{A}, \overline{B})$ measure the distance between the two continuous fuzzy numbers $\overline{A}$ and $\overline{B}$ with $D(.,.)$ given by equation(3.80) in Section 3.7. Define crisp relation R on $\mathcal{N}_0 \times \mathcal{N}_0$ as $R(\overline{A}, \overline{B}) = 1$ if and only if $D(\overline{A}, \overline{O}) = D(\overline{B}, \overline{O})$. Is R an equivalence relation? If so, describe its equivalence classes.

11. Let $D(.,.)$ be the distance metric defined in equation (3.80) in Section 3.7. Define a fuzzy relation $\overline{R}$ on the collection of continuous fuzzy numbers as

$$\overline{R}(\overline{M}, \overline{N}) = \exp(-D(\overline{M}, \overline{N})).$$

    Is $\overline{R}$ a fuzzy equivalence relation?

12. Define a crisp relation R on $\mathcal{N}_0 \times \mathcal{N}_0$ as $R(\overline{M}, \overline{N})) = 1$ if and only if $\overline{M} \cap \overline{N} \neq \overline{\phi}$. Is R an equivalence relation?

13. Suppose "$\leq$ " is a partial (or total) order on Υ. See Section 4.5 of Chapter 4. Define a crisp relation R on Υ as $R(\overline{M}, \overline{N}) = 1$ if and only if $\overline{M} \leq \overline{N}$. Is R an equivalence relation?

14. Let "$<$" and "$\approx$" be defined as in Example 4.5.1. Define R on Υ as $R(\overline{M}, \overline{N}) = 1$ if an only if $\overline{M} \approx \overline{N}$. Is R an equivalence relation?

15. Repeat Problem 10 using $D(.,.)$ from equation (3.81) in Section 3.7.

16. Repeat Problem 11 using $D(.,.)$ from equation (3.81) in Section 3.7.

17. If $\overline{R}$ is a fuzzy equivalence relation, then is $\overline{R}^{-1}$ also a fuzzy equivalence relation?

18. Show that equation (2.17) in Section 2.3 defines an equivalence relation on the set of all formulas of propositional logic.

19. Show that equation (2.21) in Section 2.3 is correct.

20. Define crisp relation R on Υ as $R(\overline{M}, \overline{N}) = 1$ if and only if $defuzz(\overline{M}) = defuzz(\overline{N})$ for the centroid defuzzifier in Section 4.6. Is R an equivalence relation on Υ?

7.5 Fuzzy Relation Equations

In this section $\overline{R}$ is a fuzzy relation on $X \times Y$, $\overline{S}$ is a fuzzy relation on $Y \times Z$ and $\overline{T}$ is a fuzzy relation on $X \times Z$. X, Y and Z are all finite sets with X having m members, Y having n elements and Z having p members. $\overline{R}$, $\overline{S}$ and $\overline{T}$ can all be represented as matrices whose members are numbers in the interval $[0,1]$. Let $\overline{R} = [r_{ij}]$ an $m \times n$ matrix, $\overline{S} = [s_{jl}]$ an $n \times p$ matrix and $\overline{T} = [t_{il}]$ an $m \times p$ matrix.

In Chapter 5 we solved simple fuzzy equations, in Chapter 6 we solved elementary fuzzy inequalities and now we wish to solve

$$\overline{R} \circ \overline{S} = \overline{T}, \tag{7.43}$$

for $\overline{R}$ given $\overline{S}$ and $\overline{T}$. This equation (7.43) is called a fuzzy relational equation. The other situation is to solve for $\overline{S}$ given $\overline{R}$ and $\overline{T}$. But this case is already covered by solving for $\overline{R}$ in equation (7.43) because we could solve

$$\overline{S}^{-1} \circ \overline{R}^{-1} = \overline{T}^{-1} \tag{7.44}$$

for $\overline{S}^{-1}$ giving $\overline{S} = (\overline{S}^{-1})^{-1}$. As usual, within the text, we will use t-norm T_m for the composition of fuzzy relations. Therefore, $\overline{R} \circ \overline{S}$ is done using the max-min composition of the two matrices $[r_{ij}]$ and $[s_{jl}]$ (see equation(7.8)).

Equation (7.43) becomes

$$\max_j(\min(r_{ij}, s_{jl})) = t_{il}, \tag{7.45}$$

for $1 \leq i \leq m$ and $1 \leq l \leq p$. We may write equation (7.45) as m separate equations. For each i, $1 \leq i \leq m$, we have

$$\max_j(\min(r_{ij}, s_{jl})) = t_{il}, \tag{7.46}$$

for $1 \leq l \leq p$.

We solve equation (7.46) for each row in $\overline{R}$. To show this another way, let $r_i = (r_{i1}, \cdots, r_{in})$ be the ith row of $\overline{R}$ and let $t_i = (t_{i1}, \cdots, t_{ip})$ be the ith row of $\overline{T}$. Then equation (7.46) is

$$r_i \circ \overline{S} = t_i, \tag{7.47}$$

for $1 \leq i \leq m$.

Let

$$S(t_i) = \{r_i | r_i \circ \overline{S} = t_i\}, \tag{7.48}$$

for $1 \leq i \leq m$. $S(t_i)$ is the set of solutions to equation (7.47). The rows in $S(t_i)$ are used to build $\overline{R}$ a solution to equation (7.43).

Let us first exclude a situation, which is easy to check, where there is no solution for $\overline{R}$, or $S(t_i) = \phi$ for at least one value of i. If, for some l,

$$max_j s_{jl} < max_i t_{il}, \tag{7.49}$$

then there is no solution for $\overline{R}$. We illustrate this result through the following example.

Example 7.5.1

Let

$$\overline{S} = \begin{pmatrix} 0.7 & 0.5 \\ 0.9 & 0.6 \end{pmatrix} \tag{7.50}$$

and

$$\overline{T} = \begin{pmatrix} 0.4 & 0.7 \\ 0.8 & 0.5 \end{pmatrix} \tag{7.51}$$

and $\overline{R}$ is a 2×2 matrix.

We find

$$\max_j s_{j1} = 0.9 > 0.8 = \max_i t_{i1}, \tag{7.52}$$

$$\max_j s_{j2} = 0.6 < 0.7 = \max_i t_{i2}. \tag{7.53}$$

So $\overline{R} \circ \overline{S} = \overline{T}$ has no solution for $\overline{R}$. Look at r_1 composed with the second column of $\overline{S}$ giving

$$\max(\min(r_{11}, 0.5), \min(r_{12}, 0.6)) < 0.7, \tag{7.54}$$

because $\min(r_{11}, 0.5) \leq 0.5$ and $\min(r_{12}, 0.6) \leq 0.6$, so the largest of the left hand side of equation (7.54) can be is 0.6. No value of r_{11} and r_{12} can make it equal to $0.7 = t_{12}$.

However, if

$$\max_j s_{jl} \geq \max_i t_{il}, \tag{7.55}$$

for all l, then there may still be no solution for $\overline{R}$.

To explain the structure of $S(t_i)$, when it is non-empty, we need to define an ordering on $\mathbf{R}^n$. If $u = (u_1, \cdots, u_n)$ and $v = (v_1, \cdots, v_n)$ are in $\mathbf{R}^n$ we write $u \leq v$ if and only if $u_i \leq v_i$, $1 \leq i \leq n$. Given $u, v \in \mathbf{R}^n$ so that $u \leq v$, then define

$$[u, v] = \{w \in \mathbf{R}^n | u \leq w \leq v\}. \tag{7.56}$$

An r_i^{**} in $S(t_i)$ is called a maximum solution if for any $r_i \in S(t_i)$ we have $r_i \leq r_i^{**}$. An r_i^* in $S(t_i)$ is called a minimal solution if for any $r_i \in S(t_i)$ if $r_i \leq r_i^*$, then $r_i = r_i^*$. It is known that if $S(t_i)$ is non-empty, then r_i^{**} is unique (only one maximum solution) but there may be more than one minimal solution. We may then describe $S(t_i)$ as

$$S(t_i) = \bigcup \{[r_i^*, r_i^{**}] | r_i^* \textit{ is a minimal solution}\}. \tag{7.57}$$

There are algorithms for both r_i^{**} and r_i^*. We shall next present the algorithm for r_i^{**} but we will omit the one for r_i^* since it is quite complicated. Instead of using an algorithm to find the r_i^* we will illustrate, through two elementary examples, how to obtain all minimal solutions. Remember that equation (7.57) only gives all solutions for each row in $\overline{R}$.

The algorithm for r_i^{**} is to first define, for fixed values of i and j,

$$\theta(s_{jk}, t_{ik}) = \begin{cases} t_{ik}, & if\ s_{jk} > t_{ik}, \\ 1, & \text{otherwise}, \end{cases} \tag{7.58}$$

for $1 \leq k \leq p$. Then if $r_i^{**} = (r_{i1}^{**}, \cdots, r_{in}^{**})$, we have

$$r_{ij}^{**} = \min_{1 \leq k \leq p} [\theta(s_{jk}, t_{ik})], \tag{7.59}$$

for $1 \leq j \leq n$. It is known that $S(t_i)$ is non-empty if and only if r_i^{**} is in $S(t_i)$, or r_i^{**} solves the equation (7.47).

Example 7.5.2

Assume $n = m = p = 4$ with

$$\overline{S} = \begin{pmatrix} 0.4 & 1 & 0.1 & 0.5 \\ 0.7 & 0 & 0.9 & 0.2 \\ 1 & 0 & 0.8 & 0.5 \\ 0.3 & 0 & 0.1 & 0.6 \end{pmatrix} \tag{7.60}$$

and $t_2 = (0.7, 0, 0.8, 0.5)$. We are to find all solutions for r_2, the second row in $\overline{R}$.

We first find r_2^{**} to see if it satisfies $r_2^{**} \circ \overline{S} = t_2$. Compute

$$r_{21}^{**} = \min\{1, 0, 1, 1\} = 0, \tag{7.61}$$

$$r_{22}^{**} = \min\{1, 1, 0.8, 1\} = 0.8, \tag{7.62}$$

$$r_{23}^{**} = \min\{0.7, 1, 1, 1\} = 0.7, \tag{7.63}$$

$$r_{24}^{**} = \min\{1, 1, 1, 0.5\} = 0.5. \tag{7.64}$$

Hence, $r_2^{**} = (0, 0.8, 0.7, 0.5)$ which satisfies $r_2^{**} \circ \overline{S} = t_2$ so the $S(t_2)$ is non-empty.

Now for the values of r_2^*. First look at r_2 and the second column of $\overline{S}$. We get $r_{21} = 0 = t_{22}$. Next consider r_2 and the third column of $\overline{S}$. Using $r_{21} = 0$ we get

$$\max(0, \min(r_{22}, 0.9), \min(r_{23}, 0.8), \min(r_{24}, 0.1)) = 0.8. \tag{7.65}$$

We must have $r_{22} \leq 0.8$. If $r_{22} < 0.8$, then $r_{23} \geq 0.8$. From r_2 and the first column in $\overline{S}$ we have $(r_{22} < 0.8, r_{23} \geq 0.8)$

$$max(0, \min(r_{22}, 0.7), \min(r_{23}, 1), \min(r_{24}, 0.3)) \geq 0.8 \tag{7.66}$$

which can never equal $0.7 = t_{21}$. Hence, $r_{22} = 0.8$.

Looking at r_2 and the first column in $\overline{S}$ we conclude $r_{23} \leq 0.7$ and from r_2 and the last column of $\overline{S}$ we see $r_{24} \leq 0.5$. We want minimal solutions so first set $r_{23} = 0$ which implies $r_{24} = 0.5$ and next set $r_{24} = 0$ implying $r_{23} = 0.5$. The two minimal solutions are:

$$r_{2a}^{*} = (0, 0.8, 0, 0.5), \tag{7.67}$$

and

$$r_{2b}^{*} = (0, 0.8, 0.5, 0). \tag{7.68}$$

Then we have all solutions

$$[r_{2a}^{*}, r_2^{**}] \cup [r_{2b}^{*}, r_2^{**}]. \tag{7.69}$$

Example 7.5.3

Assume $m = n = p = 3$ and

$$\overline{S} = \begin{pmatrix} 1 & 0.9 & 0.6 \\ 0.5 & 0.8 & 0.8 \\ 0.6 & 0.6 & 0.4 \end{pmatrix} \tag{7.70}$$

with $t_1 = (0.5, 0.6, 0.6)$. Find all solutions for r_1, the first row in $\overline{R}$.

Apply the algorithm to find r_1^{**} giving

$$r_{11}^{**} = \min\{0.5, 0.6, 1\} = 0.5, \tag{7.71}$$

$$r_{12}^{**} = \min\{1, 0.6, 0.6\} = 0.6, \tag{7.72}$$

$$r_{13}^{**} = \min\{0.5, 1, 1\} = 0.5. \tag{7.73}$$

So $r_1^{**} = (0.5, 0.6, 0.5)$ which belongs to $S(t_1)$.

To find all the minimal solutions first use r_1 and the first column of $\overline{S}$. We find that $r_{11} \leq 0.5$ and $r_{13} \leq 0.5$ must hold. Using this result next use r_1 and the second column in $\overline{S}$ to find that r_{12} must equal 0.6. Now $r_{11} \leq 0.5$, $r_{12} = 0.6$ and $r_{13} \leq 0.5$ is acceptable because using this and the third column of $\overline{S}$ we do obtain $0.6 = t_{13}$. For a minimal solution set $r_{11} = r_{13} = 0$ giving one $r_1^{*} = (0, 0.6, 0)$. All solutions are $[r_1^{*}, r_1^{**}]$.

If we were to solve larger fuzzy relational equations, then our method would become too involved and we would need that algorithm to find all minimal solutions. Or we might employ a genetic algorithm discussed in Chapter 15 (see Example 15.1 in Chapter 15).

7.5.1 Exercises

1. Show, by example, that $\overline{R} \circ \overline{S} = \overline{T}$ may have no solution for $\overline{R}$ when (equation 7.55)

$$\max_j s_{jl} \geq \max_i t_{il},$$

for all l.

2. Using $\overline{S}$ and t_2 from Example 7.5.2 determine if $r_2 \circ \overline{S} = t_2$ has a solution for r_2 if we use t-norm T_β for composition. If it has a solution, then find a maximum solution.

   a. $T_\beta = T_b$.
   b. $T_\beta = T_p$.
   c. $T_\beta = T^*$.

3. Using $\overline{S}$ and t_1 from Example 7.5.3 determine if $r_1 \circ \overline{S} = t_1$ has a solution for r_1 if we use t-norm T_β for composition. If it has a solution, then find a maximum solution.

   a. $T_\beta = T_b$.
   b. $T_\beta = T_p$.
   c. $T_\beta = T^*$.

4. Using t-norm T_m and

$$\overline{S} = \begin{pmatrix} 0.3 & 0.9 & 0 \\ 0 & 0.2 & 0.8 \\ 0.7 & 0 & 1 \end{pmatrix}$$

and

$$\overline{T} = \begin{pmatrix} 0.4 & 0.8 & 0.7 \\ 0.4 & 0.9 & 0.3 \\ 0.7 & 0.2 & 1 \end{pmatrix}$$

determine if $\overline{R} \circ \overline{S} = \overline{T}$ has a solution for $\overline{R}$. If it has a solution, then describe the solution sets $S(t_i)$ where t_i is the ith row in $\overline{T}$, $1 \leq i \leq 3$.

5. Using the $\overline{S}$ and $\overline{T}$ from problem 4, does $\overline{R} \circ \overline{S} = \overline{T}$ have a solution for $\overline{R}$ if we are using t-norm T_β for composition? If so, then find one solution for $\overline{R}$.

   a. $T_\beta = T_b$.
   b. $T_\beta = T_p$.
   c. $T_\beta = T^*$.

6. Using t-norm min and

$$\overline{S}=\begin{pmatrix} 0.6 & 0.5 \\ 0.9 & 0.7 \end{pmatrix}$$

and

$$\overline{T}=\begin{pmatrix} 1 & 0.8 \\ 0.4 & 0.3 \end{pmatrix}$$

does $\overline{R}\circ\overline{S}=\overline{T}$ have a solution for $\overline{R}$? If so, describe all solutions for $\overline{R}$.

7. Using t-norm min and

$$\overline{S}=\begin{pmatrix} 0.5 & 0.4 & 0.6 \\ 0.2 & 0 & 0.6 \end{pmatrix}$$

and

$$\overline{T}=\begin{pmatrix} 0.7 & 0.8 & 0.5 \\ 1 & 0.2 & 0.6 \end{pmatrix}$$

does $\overline{R}\circ\overline{S}=\overline{T}$ have a solution for $\overline{R}$? If so, describe all solutions for $\overline{R}$.

8. Using t-norm min and

$$\overline{S}=\begin{pmatrix} 0.6 & 1 \\ 0.4 & 0.9 \\ 0 & 0.7 \end{pmatrix}$$

and

$$\overline{T}=\begin{pmatrix} 0.5 & 0 \\ 0.4 & 1 \end{pmatrix}$$

does $\overline{R}\circ\overline{S}=\overline{T}$ have a solution for $\overline{R}$? If so, describe all solutions for $\overline{R}$.

Chapter 8

Fuzzy Functions

8.1 Introduction

Just as crisp function are important in mathematical modeling, fuzzy functions are important in fuzzy modeling. The usual way to obtain a fuzzy function is to extend a crisp function to map fuzzy sets to fuzzy sets, and there are two common methods to accomplish this extension. The first method, called the extension principle procedure, is discussed in the next section, and the second method, called the α-cut and interval arithmetic procedure, is presented in section three. In a pre-calculus course you study different classes of functions including linear, quadratic, polynomial, radical, exponential and logarithmic and we do this for fuzzy functions in section four. Fuzzy trigonometric functions are in Chapter 10. Also in pre-calculus you would study inverse functions and section five is about fuzzy inverse functions. Elementary differential calculus of fuzzy functions is introduced in the last section, section six.

8.2 Extension Principle

Let X and Y be two sets and let $\mathcal{F}(X)$ and $\mathcal{F}(Y)$ denote all fuzzy subsets of X and Y, respectively. A fuzzy function $\overline{F}$ is simply a function mapping $\mathcal{F}(X)$ into $\mathcal{F}(Y)$. If $\overline{A}$ is a fuzzy subset of X. then $\overline{F}(\overline{A}) = \overline{B}$ is a fuzzy subset of Y. Let $\Omega = \Omega(X;Y)$ denote all such fuzzy functions.

Example 8.2.1

Let $X = \{x_1, x_2, x_3\}$ and $Y = \{y_1, y_2\}$. $\overline{A}$, a fuzzy subset of X, is written as

$$\overline{A} = \{\frac{a_1}{x_1}, \frac{a_2}{x_2}, \frac{a_3}{x_3}\}, \tag{8.1}$$

where the membership values are the a_i. That is, $\overline{A}(x_i) = a_i$, $1 \leq i \leq 3$.

If $\overline{B}$ is a fuzzy subset of Y, then we may write $\overline{B}$ as

$$\overline{B} = \{\frac{b_1}{y_1}, \frac{b_2}{y_2}\}. \tag{8.2}$$

If we let $b_1 = \min\{a_1, a_2\}$ and $b_2 = \max\{a_2, a_3\}$, then this describes a fuzzy function.

Example 8.2.2

Let $X = Y$ be the set of real numbers and $\overline{E}_i$ fixed trapezoidal fuzzy numbers, $i = 1, 2$. Then $\overline{B} = (\overline{A} \cap \overline{E}_1) \cup (\overline{A}^c \cap \overline{E}_2)$ defines a fuzzy function $\overline{F}(\overline{A}) = \overline{B}$ for $\overline{A}$ any fuzzy subset of the real numbers.

Where do all the $\overline{F}$ in Ω come from? Many members of Ω come from extending non-fuzzy f mapping X into Y to $\overline{F}$ mapping $\mathcal{F}(X)$ into $\mathcal{F}(Y)$. This extension may be accomplished using the extension principle. Before we present the extension principle for extending f to $\overline{F}$ let us review the crisp case. We have an f mapping X into Y and we wish to extend it to $\widetilde{f}$ mapping $\mathcal{P}(X)$ into $\mathcal{P}(Y)$. $\mathcal{P}(X)$ ($\mathcal{P}(Y)$) is called the power set of X (Y) and it is the set of all subsets (not fuzzy) of X (Y). If A is a crisp subset of X, then $\widetilde{f}(A) = B$, for B being a crisp subset of Y. The usual method of obtaining B is

$$B = \{y \in Y | y = f(x), x \in X\}. \tag{8.3}$$

Recall that we write the characteristic function for a set D as $D(x)$ which equals one when x belongs to D and $D(x) = 0$ otherwise. Then we can rewrite equation (8.3) as

$$B(y) = (\widetilde{f}(A))(y) = max\{A(x) | y = f(x)\}, \tag{8.4}$$

where we interpret equation (8.4) to give value zero for $B(y)$ if there are no x in X such that $f(x) = y$. We used "max" in equation (8.4) because the $A(x)$ values are either zero or one.

Now we extend this idea in equation (8.4) to f mapping X into Y to obtain $\overline{F}$ in Ω. If $\overline{B} = \overline{F}(\overline{A})$ we find the membership values of $\overline{B}$ as

$$\overline{B}(y) = \sup\{\overline{A}(x) | f(x) = y\}. \tag{8.5}$$

Equation (8.5) gives zero for $\overline{B}(y)$ if there is no x in X that makes $f(x) = y$. We now use "sup" since the values of $\overline{A}(x)$ can range throughout $[0, 1]$.

Let Ω_e be all $\overline{F}$ in Ω which are the extension principle extension of some f mapping X into Y. Ω_e is a subset of Ω.

Table 8.1: The Function in Example 8.2.3

x	y
x_1	y_2
x_2	y_1
x_3	y_2
x_4	y_3
x_5	y_1

Example 8.2.3

Let $X = \{x_1, \cdots, x_5\}$ and $Y = \{y_1, y_2, y_3\}$. Define f mapping X into Y as shown in Table 8.1.

Next let

$$\overline{A} = \{\frac{0}{x_1}, \frac{0.3}{x_2}, \frac{1}{x_3}, \frac{0.8}{x_4}, \frac{0.5}{x_5}\}, \tag{8.6}$$

and

$$\overline{B} = \{\frac{b_1}{y_1}, \frac{b_2}{y_2}, \frac{b_3}{y_3}\}. \tag{8.7}$$

Extend f to $\overline{F}$ and then find $\overline{B}$ where $\overline{B} = \overline{F}(\overline{A})$. We calculate

$$b_1 = \overline{B}(y_1) = \max\{0.3, 0.5\} = 0.5, \tag{8.8}$$

$$b_2 = \overline{B}(y_2) = \max\{0, 1\} = 1, \tag{8.9}$$

$$b_3 = \overline{B}(y_3) = 0.8. \tag{8.10}$$

Can Ω_e be all of Ω? We now argue that there are many $\overline{F}$ in Ω which are not in Ω_e. A fuzzy function $\overline{G}$ in Ω is monotone increasing if whenever $\overline{A} \leq \overline{B}$, $\overline{A}$ and $\overline{B}$ being fuzzy subsets of X, then $\overline{G}(\overline{A}) \leq \overline{G}(\overline{B})$. Every $\overline{F}$ in Ω_e is monotone increasing. Since there are $\overline{F}$ in Ω which are not monotone increasing, we see that $\Omega \neq \Omega_e$. Let Ω_m be all $\overline{F}$ in Ω which are monotone increasing. Is Ω_m equal to Ω_e? We know that $\Omega_e \subset \Omega_m \subset \Omega$.

The above development may be generalized to functions of many independent variables. Let X, Y and Z be sets and f a mapping from $X \times Y$ into Z. That is, $z = f(x, y)$ for $x \in X$, $y \in Y$ and $z \in Z$. Also let $\mathcal{F}(Z)$ denote all fuzzy subsets of Z. Define $\Omega = \Omega(X, Y; Z)$ be all fuzzy functions $\overline{F}$ mapping $\mathcal{F}(X) \times \mathcal{F}(Y)$ into $\mathcal{F}(Z)$. So if $\overline{A}$ ($\overline{B}$) is a fuzzy subset of $X(Y)$, then $\overline{F}(\overline{A}, \overline{B}) = \overline{C}$ for $\overline{C}$ being a fuzzy subset of Z. The extension principle may be used to extend f to $\overline{F}$ in Ω. The membership function for $\overline{C}$ is then defined by

$$\overline{C}(z) = \sup\{\min(\overline{A}(x), \overline{B}(y)) | f(x, y) = z\}, \tag{8.11}$$

with $\overline{C}(z) = 0$ if there is no $(x, y) \in X \times Y$ that will produce $f(x, y) = z$. This is the extension principle applied to f. Notice that we used t-norm T_m in equation (8.11). We could have used T_b, T_p or T^* and probably get a different value for $\overline{C}$. Let Ω_e be all $\overline{F}$ in Ω coming from the extension principle. Each $\overline{F}$ in Ω_e is monotone increasing which now means that if $\overline{A}_1 \leq \overline{A}_2$ and $\overline{B}_1 \leq \overline{B}_2$, then $\overline{F}(\overline{A}_1, \overline{B}_1) \leq \overline{F}(\overline{A}_2, \overline{B}_2)$.

Example 8.2.4

Let $X = Y = Z$ be the real number interval $(0, \infty)$ and define

$$z = f(x, y) = \frac{xy}{x + y}. \tag{8.12}$$

The extension principle will give values for $\overline{C}$ where

$$\overline{C} = \frac{\overline{A} \cdot \overline{B}}{\overline{A} + \overline{B}}, \tag{8.13}$$

for $\overline{A}$ and $\overline{B}$ fuzzy subsets of $(0, \infty)$.

For the rest of this section we will restrict our attention to $X = Y = Z$ the set of real numbers. Also, let us only use continuous fuzzy numbers, as fuzzy subsets of the real numbers, that will be inputs to our fuzzy functions. f will be a function mapping D, an interval in $\mathbf{R}$, into the real numbers. We will extend f, via the extension principle, to fuzzy function $\overline{F}$ mapping $\overline{A}$, a continuous fuzzy number in D, to $\overline{B}$ a fuzzy subset of $\mathbf{R}$.

If f is continuous, then we have a way to find α-cuts of $\overline{B} = \overline{F}(\overline{A})$. Let $\overline{B} = [b_1(\alpha), b_2(\alpha)]$, since α-cuts of $\overline{B}$ will be closed intervals when f is continuous. Then

$$b_1(\alpha) = \min\{f(x) | x \in \overline{A}[\alpha]\}, \tag{8.14}$$

$$b_2(\alpha) = \max\{f(x) | x \in \overline{A}[\alpha]\}, \tag{8.15}$$

for $0 \leq \alpha \leq 1$. We may use min and max in equations (8.14) and (8.15), respectively, because a continuous function on a closed interval $\overline{A}[\alpha]$ takes on its max and min.

Equations (8.14) and (8.15) can give us an easy way to find α-cuts of $\overline{B} = \overline{F}(\overline{A})$. Suppose f is monotonically increasing on D. Then we see that $b_1(\alpha) = f(a_1(\alpha))$ and $b_2(\alpha) = f(a_2(\alpha))$ where $\overline{A}[\alpha] = [a_1(\alpha), a_2(\alpha)]$. This is true for $y = f(x) = e^x$, $tan(x)$ on $(-\pi/2, \pi/2)$ and $\ln(x)$ for $x > 0$, etc.

Now we can generalize to f mapping $D_1 \times D_2$, D_1 and D_2 intervals in $\mathbf{R}$, into the real numbers. Assume that f is a continuous function and using the extension principle extend f to $\overline{F}$ where $\overline{C} = \overline{F}(\overline{A}, \overline{B})$, $\overline{A}$ ($\overline{B}$) being a continuous fuzzy number in D_1 (D_2). If $\overline{C}[\alpha] = [c_1(\alpha), c_2(\alpha)]$, then

$$c_1(\alpha) = \min\{f(x,y)|x \in \overline{A}[\alpha], y \in \overline{B}[\alpha]\}, \tag{8.16}$$

$$c_2(\alpha) = \max\{f(x,y)|x \in \overline{A}[\alpha], y \in \overline{B}[\alpha]\}, \tag{8.17}$$

for $0 \leq \alpha \leq 1$. We see that $\overline{C}$ depends on which t-norm we are using in the extension principle. We have been using T_m so equations (8.16) and (8.17) are only known to hold for t-norm min. Suppose $\partial f/\partial x > 0$ and $\partial f/\partial y < 0$. Then we see from equations (8.16) and (8.17) that $c_1(\alpha) = f(a_1(\alpha), b_2(\alpha))$ and $c_2(\alpha) = f(a_2(\alpha), b_1(\alpha))$ for all α.

Example 8.2.5

Let

$$z = f(x,y) = \frac{2x - y}{x + y}, \tag{8.18}$$

for $x, y \in (0, \infty)$. Let us find α-cuts of $\overline{C}$ where

$$\overline{C} = \frac{2\overline{A} - \overline{B}}{\overline{A} + \overline{B}}, \tag{8.19}$$

for $\overline{A}, \overline{B}$ continuous fuzzy numbers in $(0, \infty)$. Since $\partial f/\partial x > 0$ and $\partial f/\partial y < 0$ we obtain

$$c_1(\alpha) = \frac{2a_1(\alpha) - b_2(\alpha)}{a_1(\alpha) + b_2(\alpha)}, \tag{8.20}$$

$$c_2(\alpha) = \frac{2a_2(\alpha) - b_1(\alpha)}{a_2(\alpha) + b_1(\alpha)}, \tag{8.21}$$

for $0 \leq \alpha \leq 1$. $\overline{C}$ in equation (8.19) is obtained using the extension principle, and using t-norm min, and the α-cuts of $\overline{C}$ are shown in equations (8.20) and (8.21) for $\overline{B}[\alpha] = [b_1(\alpha), b_2(\alpha)]$, $\overline{A}[\alpha] = [a_1(\alpha), a_2(\alpha)]$.

Fuzzy expressions may also be evaluated using the extension principle. For example

$$\overline{B} = \overline{F}(\overline{A}) = \frac{2\overline{A} + 10}{3\overline{A} - 4}, \tag{8.22}$$

would be the extension of

$$y = f(x) = \frac{2x + 10}{3x - 4}. \tag{8.23}$$

The fuzzy expression

$$\overline{E} = \overline{F}(\overline{A}, \overline{B}, \overline{C}, \overline{D}, \overline{W}) = \frac{\overline{A} \cdot \overline{W} + \overline{B}}{\overline{C} \cdot \overline{W} + \overline{D}}, \tag{8.24}$$

would be the extension of

$$z = f(x_1, x_2, x_3, x_4, x_5) = \frac{x_1 x_5 + x_2}{x_3 x_5 + x_4}. \tag{8.25}$$

The extension giving equation (8.24) may use any t-norm. If we use $T_m = \min$, then let

$$\Pi(x_1, x_2, x_3, x_4, x_5) = \min\{\overline{A}(x_1), \cdots, \overline{W}(x_5)\}. \tag{8.26}$$

Then

$$\overline{E}(z) = \sup\{\Pi(x_1, \cdots, x_5) | f(x_1, \cdots, x_5) = z\}. \tag{8.27}$$

Of course we could use expressions like equations (8.16) and (8.17) to find α-cuts of $\overline{E}$. Other t-norms may be used in equation (8.26) to get Π.

8.2.1 Exercises

1. Show that equation (8.4) is correct.

2. Let f map X into Y and let $\overline{F}$ be the extension principle extension of f. If $\overline{A}$ and $\overline{A}_i$, $1 \leq i \leq n$ are fuzzy subsets of X, then determine whether or not the following equations are true or false:

   a. $\overline{F}(\bigcup_{i=1}^{n} \overline{A}_i) = \bigcup_{i=1}^{n} \overline{F}(\overline{A}_i)$,

   b. $\overline{F}(\bigcap_{i=1}^{n} \overline{A}_i) = \bigcap_{i=1}^{n} \overline{F}(\overline{A}_i)$, and

   c. $f(\overline{A}[\alpha]) = \overline{F}(\overline{A})[\alpha], 0 \leq \alpha \leq 1$.

3. Show that if $\overline{F}$ is the extension principle extension of some f mapping X into Y, then $\overline{F}$ must be monotone increasing.

4. Find X, Y and $\overline{F}$ in Ω so that $\overline{F}$ is not monotone increasing.

5. Determine whether or not Ω_m and Ω_e are equal. If $\Omega_m \neq \Omega_e$, then find a X, Y and $\overline{F}$ in Ω_m so that $\overline{F}$ is not in Ω_e.

6. Let f map $[0,1]$ into $[0,1]$ given by the expression

$$f(x) = \begin{cases} x, & 0 \leq x \leq 0.4 \\ 0.4, & 0.4 \leq x \leq 0.6 \\ 1-x, & 0.6 \leq x \leq 1 \end{cases}$$

   Let $\overline{A} = (\frac{1}{3}/0.5/\frac{2}{3})$. $\overline{F}$ is the extension principle extension of f. Let $\overline{B} = \overline{F}(\overline{A})$.

   a. Find $\overline{B}$.

   b. Show that $\overline{B}$ is not a continuous fuzzy number.

   c. Find the α-cuts of $\overline{B}$.

   This example shows that $\overline{B} = \overline{F}(\overline{A})$ need not be a continuous fuzzy number even if $\overline{A}$ is a continuous fuzzy number and f is a continuous function.

7. Let $X = \{k(\pi/6) | k = 0, \pm1, \cdots, \pm12\}$, $Y = \mathbf{R}$, $f(x) = \sin(x)$ for x in X and $\overline{F}$ the extension principle extension of f. If $\overline{A}$ is a fuzzy subset of X, then find $\overline{B} = \overline{F}(\overline{A})$.

8. Using equation (8.11) show that $\overline{F}$ must be monotone increasing if the t-norm is:

   a. T_m,

   b. T_b,

c. T_p, and

d. T^*.

9. Let $\overline{A} = (1/3, 5/8)$ and $\overline{B} = (18/20/24)$. Find α-cuts of $\overline{C}$ in Example 8.2.4 using t-norm:

   a. T_m,

   b. T_b,

   c. T_p, and

   d. T^*.

10. Find α-cuts of $\overline{C}$ in Example 8.2.5 using t-norm:

    a. T_b,

    b. T_p, and

    c. T^*.

11. As a result of Problems 9 and 10 what can be said, if anything, about the relationship between the $\overline{C}$ values, $\overline{C} = \overline{F}(\overline{A})$, using the four t-norms: T_m,T_b,T_p, and T^*.

12. Let $f : X \to Y$, $X = \{x_1, \cdots, x_n\}$, $Y = \{y_1, \cdots, y_m\}$, $n \leq m$ and assume that f is one-to-one. Extend f to $\overline{F}$ from the extension principle. If

$$\overline{A} = \{\frac{\mu_1}{x_1}, \cdots, \frac{\mu_n}{x_n}\},$$

and $\overline{B} = \overline{F}(\overline{A})$, then show that

$$\overline{B} = \{\frac{\mu_1}{f(x_1)}, \cdots, \frac{\mu_n}{f(x_n)}\},$$

with all other values y in Y having membership value zero.

13. Let f be a continuous function mapping an interval $[a, b]$ in $\mathbf{R}$ into the real numbers. $\overline{F}$ is the extension principle extension of f and $\overline{A}$ is a continuous fuzzy number in $[a, b]$. Determine if the following equation is true or false:

    a. $f(co(\overline{A})) = co(\overline{F}(\overline{A}))$, where "$co$" is core,

    b. $f(base(\overline{A})) = base(\overline{F}(\overline{A}))$, where "$base$" is the α-cut, $\alpha = 0$.

14. Let $y = f(x) = \sin(x)$ and $\overline{A} = (0/\pi/2\pi)$. Find $\overline{F}(\overline{A})$.

15. Fuzzy $\overline{\max}$ and $\overline{\min}$, see Section 4.4, are fuzzy functions mapping pairs of fuzzy numbers into fuzzy subsets of $\mathbf{R}$. Are they monotone increasing fuzzy functions?

16. Determine if equations (8.16) and (8.17) remain true if we use t-norm:

    a. T_b,

    b. T_p, and

    c. T^*.

17. Suppose $f : \mathbf{R} \rightarrow \mathbf{R}$ is one-to-one. Let $\overline{F}$ be its extension principle extension. Give an "easy" way to find $\overline{C} = \overline{F}(\overline{A})$; $\overline{A}$ is a fuzzy subset of the real numbers.

18. Let $z = f(x, y)$ be continuous from $(x, y) \in [a, b] \times [c, d]$ into the real numbers. $\overline{F}$ is the extension principle extension of f. Let $\overline{C} = \overline{F}(\overline{A}, \overline{B})$ for continuous fuzzy number $\overline{A}$ ($\overline{B}$) in $[a, b]$ ($[c, d]$). We will measure the fuzziness of $\overline{C}$ by the length of its base ($\overline{C}[0]$). Using t-norms T_m,T_b,T_p, and T^* which, if any, will always minimize the fuzziness of $\overline{C}$?

19. Find α-cuts of $e^{\overline{A}}$ for any continuous fuzzy number $\overline{A}$.

20. How would you define a fuzzy function of type 2 (level 2) fuzzy set (see Section 3.2).

21. In crisp mathematics a function is defined as a special type of relation (each x in the domain of the function corresponds to exactly one y in the range of the function). Is a fuzzy function a special type of fuzzy relation? Explain.

8.3 Alpha-Cuts and Interval Arithmetic

In this section all our universal sets $(X, Y, Z, \cdots)$ will be the set of real numbers, f is a function on D, an interval in $\mathbf{R}$, with values in $\mathbf{R}$. We will only be using continuous fuzzy numbers as input to our fuzzy functions. In this case there is another way to extend f to $\overline{F}$, where $\overline{F}(\overline{A}) = \overline{B}$, $\overline{B}$ a fuzzy subset of $\mathbf{R}$ and $\overline{A}$ a continuous fuzzy number in D.

For all the functions we usually use in engineering and science we have an algorithm, using a finite number of additions, subtractions, multiplications and divisions, to compute the function to desired accuracy. Such functions can be extended, using α-cuts and interval arithmetic, to fuzzy functions. We will usually be considering only those crisp functions where the algorithm provides exact values with no approximation. If the algorithm does not provide exact values for the function, then we explicitly point out that we are using an approximation. Let $y = f(x)$ be such a function. We compute α-cuts of $\overline{B} = \overline{F}(\overline{A})$ as

$$[b_1(\alpha), b_2(\alpha)] = f([a_1(\alpha), a_2(\alpha)]). \tag{8.28}$$

We input the interval $[a_1(\alpha), a_2(\alpha)] = \overline{A}[\alpha]$ into f, perform the arithmetic operations needed to evaluate f on this interval, and obtain the interval $[b_1(\alpha), b_2(\alpha)] = \overline{B}[\alpha]$. Notice that $f(\overline{A}[\alpha])$ in equation (8.28) does not necessarily mean $\{y | y = f(x), x \in \overline{A}[\alpha]\}$.

Example 8.3.1

Let $y = f(x) = 2x^2 - 3x + 10$ for x in $\mathbf{R}$. Extend f to $\overline{F}$ using α-cuts and interval arithmetic. Let $\overline{A} \approx (a_1/a_2, a_3/a_4)$ with $\overline{A}[\alpha] = [a_1(\alpha), a_2(\alpha)]$. Then if $\overline{B} = \overline{F}(\overline{A})$, we have

$$\overline{B}[\alpha] = 2[a_1(\alpha), a_2(\alpha)]^2 - 3[a_1(\alpha), a_2(\alpha)] + 10. \tag{8.29}$$

Let us assume that $\overline{A} > 0$, or $a_1 > 0$, so that

$$[a_1(\alpha), a_2(\alpha)]^2 = [a_1^2(\alpha), a_2^2(\alpha)]. \tag{8.30}$$

Then

$$\overline{B}[\alpha] = [2a_1^2(\alpha) - 3a_2(\alpha) + 10, 2a_2^2(\alpha) - 3a_1(\alpha) + 10], \tag{8.31}$$

for $0 \leq \alpha \leq 1$, If $a_4 < 0$, or $a_1 < 0 < a_4$, then we would obtain a different result. We used interval arithmetic discussed in Section 4.3.2 to evaluate $f(\overline{A}[\alpha])$.

If $z = f(x, y)$ is also one of these usual functions of science and engineering, we can extend f to $\overline{F}$ via α-cuts and interval arithmetic. Let

$\overline{C} = \overline{F}(\overline{A}, \overline{B})$, $\overline{A}$ ($\overline{B}$) a continuous fuzzy number in interval D_1 (D_2), then

$$\overline{C}[\alpha] = f(\overline{A}[\alpha], \overline{B}[\alpha]), \tag{8.32}$$

for $0 \leq \alpha \leq 1$.

Example 8.3.2

This example continues Example 8.2.5 with $z = f(x, y) = (2x - y)/(x + y)$, for $x, y \in (0, \infty)$. Extend f to $\overline{F}$ using the method of this section and set $\overline{C} = \overline{F}(\overline{A}, \overline{B})$. If $\overline{A}[\alpha] = [a_1(\alpha), a_2(\alpha)]$, $\overline{B}[\alpha] = [b_1(\alpha), b_2(\alpha)]$, then

$$\overline{C}[\alpha] = \frac{2[a_1(\alpha), a_2(\alpha)] - [b_1(\alpha), b_2(\alpha)]}{[a_1(\alpha), a_2(\alpha)] + [b_1(\alpha), b_2(\alpha)]}, \tag{8.33}$$

or

$$\overline{C}[\alpha] = \frac{[2a_1(\alpha) - b_2(\alpha), 2a_2(\alpha) - b_1(\alpha)]}{[a_1(\alpha) + b_1(\alpha), a_2(\alpha) + b_2(\alpha)]}, \tag{8.34}$$

so that

$$c_1(\alpha) = \frac{2a_1(\alpha) - b_2(\alpha)}{a_2(\alpha) + b_2(\alpha)}, \tag{8.35}$$

and

$$c_2(\alpha) = \frac{2a_2(\alpha) - b_1(\alpha)}{a_1(\alpha) + b_1(\alpha)}, \tag{8.36}$$

for $0 \leq \alpha \leq 1$. The above interval arithmetic computations are correct since $\overline{A}$ and $\overline{B}$ are in $(0, \infty)$, or $\overline{A} > 0$ and $\overline{B} > 0$.

Fuzzy expressions, equations (8.22) and (8.24) may also be evaluated using α-cuts and interval arithmetic.

Now let us compare these two methods of extending f to $\overline{F}$. If $\overline{F}$ comes from the extension principle (using t-norm T_m) let us denote $\overline{F}$ as $\overline{F}_e$ and write $\overline{F}$ as $\overline{F}_i$ if we employ α-cuts and interval arithmetic. For all the usual functions of engineering and science we get

$$\overline{F}_e(\overline{A}) \leq \overline{F}_i(\overline{A}), \tag{8.37}$$

for all continuous fuzzy numbers $\overline{A}$ in D. This result also holds for functions of more than one independent variable ($z = f(x_1, \cdots, x_n)$,$n \geq 2$).

Although we obtain the same results in elementary arithmetic using the extension principle (using t-norm T_m), or α-cuts and interval arithmetic (Section 4.3.3), for more complicated fuzzy expressions the two procedures can produce different results.

Example 8.3.3

Let

$$y = f(x) = x(1 - x), \tag{8.38}$$

for $0 \leq x \leq 1$. Let $\overline{A} = (0/0.25/0.5)$ and set $\overline{B} = \overline{F}(\overline{A})$. Using $\overline{F}_e$ write $\overline{B}$ as $\overline{B}_e$ and write $\overline{B}_i$ for $\overline{B}$ if it is obtained from $\overline{F}_i$. We compute $\overline{B}_e[0.5]$ and $\overline{B}_i[0.5]$ and we find that they are not equal (see Exercises).

There is no known necessary and sufficient condition on $y = f(x)$ (or $y = f(x, y), \cdots$) so that $\overline{F}_e = \overline{F}_i$. What we do know is that if each fuzzy number appears only once in the expression (and is not squared, etc.), then the two methods are expected to produce the same results ($\overline{F}_e = \overline{F}_i$).

So, when evaluating fuzzy functions (input-output fuzzy numbers) you always need to tell which method you are using. We expect the α-cut and interval arithmetic procedure to produce a more fuzzy result ($\overline{F}_e \leq \overline{F}_i$). Our policy will be: (1) always try the extension principle method first, and and (2) if it is too difficult to evaluate $\overline{F}_e$, see equations (8.14)–(8.17), then use $\overline{F}_i$ as an approximation to $\overline{F}_e$.

8.3.1 Exercises

1. In Example 8.3.1, is $\overline{B}[\alpha]$ given in equation (8.31) the same as

$$\{y|f(x) = y, x \in \overline{A}[\alpha]\}?$$

2. Let $\overline{C}_e[\alpha]$ be the value of $\overline{C}[\alpha]$ from Example 8.2.5 and let $\overline{C}_i[\alpha]$ be the value of $\overline{C}[\alpha]$ in Example 8.3.2. Show $\overline{C}_e[\alpha] \subseteq \overline{C}_i[\alpha]$ for all α.

3. Let f be a rational function, or the quotient of two polynomials, for x in interval D. Show $\overline{F}_e(\overline{A}) \leq \overline{F}_i(\overline{A})$ for all continuous fuzzy numbers $\overline{A}$ in D.

4. Let f map D, an interval in $\mathbf{R}$, into the real numbers. Find necessary and sufficient conditions on f so that $\overline{F}_e = \overline{F}_i$.

5. Find $\overline{B}[\alpha]$ in Example 8.3.1 if $\overline{A} < 0$.

6. Find $\overline{C}[\alpha]$ in Example 8.3.2 if:

   a. $\overline{A} < 0$ and $\overline{B} < 0$,

   b. $\overline{A} < 0$ and $\overline{B} > 0$,

   c. $\overline{A} > 0$ and $\overline{B} < 0$.

7. In Example 8.3.3 find :

   a. $\overline{B}_e[0]$ and $\overline{B}_i[0]$,

   b. $\overline{B}_e[0.5]$ and $\overline{B}_i[0.5]$.

8. Let $z = f(x, y) = (x + y)/x$ for $x > 0$. Does $\overline{F}_e = \overline{F}_i$? Try $\overline{A} = \overline{B} = (0/1/2)$ and compute $\overline{C} = \overline{F}(\overline{A}, \overline{B})$ both ways. Use t-norm $T_m = \min$.

9. Let $z = f(x, y) = x + y^2$ for x and y real numbers. Does $\overline{F}_e = \overline{F}_i$? Try $\overline{A} = \overline{B} = (-1/0/1)$ and find $\overline{C} = \overline{F}(\overline{A}, \overline{B})$ using both procedures. Use t-norm T_m.

10. Redo Problem 8 using t-norm T_p.

11. Redo problem 9 using t-norm T_p.

12. Let $f : D \to \mathbf{R}$, D an interval in $\mathbf{R}$ and assume that f is continuous. $\overline{F}_i$ is the fuzzy extension using α-cuts and interval arithmetic. Is $\overline{F}_i$ monotone increasing?

13. In the following problems you are given $y = f(x)$ and D, $x \in D$ an interval in the set of real numbers. Determine if $\overline{F}_e = \overline{F}_i$ or $\overline{F}_e \neq \overline{F}_i$. If they are not equal, then find a triangular fuzzy number $\overline{A}$ in the domain of f and a value of $\alpha \in [0, 1]$ so the $\overline{F}_e[\alpha] \neq \overline{F}_i[\alpha]$.

1. $y = f(x) = (x+1)/x$, $x > 0$.
2. $y = f(x) = (x+1)/x$, $-1 < x < 0$.
3. $y = f(x) = 10x^2 - x + 1$, $x \geq 1$.
4. $y = f(x) = 10x^2 + x + 1$, $x \geq 1$.
5. $y = f(x) = x^2 - x$, $0 < x < 1$.
6. $y = f(x) = x^2 + x$, $0 < x < 1$.
7. $y = f(x) = xsin(x)$, $0 < x < \pi/2$.
8. $y = f(x) = xsin(x)$, $-\pi/2 < x < 0$.
9. $y = f(x) = xe^x$, $x > 1$.
10. $y = f(x) = e^x/x$, $x > 1$.
11. $y = f(x) = x\ln(x)$, $x \geq 3$.
12. $y = f(x) = \ln(x)/x$, $x \geq 3$.

8.4 Types of Fuzzy Functions

In this section all fuzzy sets will be fuzzy subsets of the real numbers and inputs to all fuzzy functions will be continuous fuzzy numbers. We will be looking at various types of elementary fuzzy functions up to fuzzy trigonometric functions. Fuzzy trigonometry is discussed in Chapter 10.

We start off with the fuzzy linear function which fuzzifies $y = f(x) = ax + b$. We can: (1) make x fuzzy and have a and b crisp; (2) make a and b fuzzy and have x crisp; or (3) make a, b and x fuzzy (the fully fuzzified linear function). We again will use subscript "e" if the extension principle is used (using t-norm T_m) and subscript "i" when α-cuts and interval arithmetic were used to fuzzify the crisp equation.

The fully fuzzified linear function is

$$\overline{Y} = \overline{F}(\overline{X}) = \overline{A} \cdot \overline{X} + \overline{B}, \tag{8.39}$$

where $\overline{X}, \overline{Y}, \overline{A}, \overline{B}$ are all continuous fuzzy numbers. If $\overline{X} \neq \overline{A}$, then $\overline{Y}_e = \overline{Y}_i$, but if $\overline{X} = \overline{A}$, then $\overline{Y} = \overline{A}^2 + \overline{B}$ and (see problem 9, section 8.3.1) $\overline{Y}_e$ may not equal $\overline{Y}_i$.

Example 8.4.1

In this example we only fuzzify a and b giving the fuzzy linear function $\overline{Y} = \overline{F}(x) = \overline{A}x + \overline{B}$ for crisp x.

In measuring the increasing trend in atmospheric concentrations of carbon dioxide in parts per million (ppm) from 1965 to 1998 it was found that in 1965 it was approximately 315 ppm and the measurement in 1998 was about 365 ppm. We model these two measurements as fuzzy numbers $\overline{Y}_0 = (285/300, 330/345)$ for 1965 and $\overline{Y}_1 = (330/350, 380/400)$ for 1998. We want to explain this data using

$$\overline{Y} = \overline{F}(x) = \overline{A}x + \overline{B}, \tag{8.40}$$

where $x = 0$ corresponds to 1965 and x=33 denotes 1998. Find $\overline{A}$ and $\overline{B}$ so that $\overline{F}(0) = \overline{Y}_0$ and $\overline{F}(33) = \overline{Y}_1$. Then predict $\overline{Y}$ for 2001 (x=36).

We start with $x = 0$ so that

$$\overline{A}0 + \overline{B} = \overline{Y}_0, \tag{8.41}$$

and $\overline{B} = \overline{Y}_0$. Now set $x = 33$ and solve for $\overline{A}$ using α-cuts and interval arithmetic. Let $\overline{Y}_0 = [y_{01}(\alpha), y_{02}(\alpha)]$ and $\overline{Y}_1 = [y_{11}(\alpha), y_{12}(\alpha)]$. Then if $[a_1(\alpha), a_2(\alpha)] = \overline{A}[\alpha]$, we get

$$a_1(\alpha) = (1/33)(y_{11}(\alpha) - y_{01}(\alpha)), \tag{8.42}$$

or

$$a_1(\alpha) = (1/33)(45 + 5\alpha), \tag{8.43}$$

and

$$a_2(\alpha) = (1/33)(y_{12}(\alpha) - y_{02}(\alpha)), \tag{8.44}$$

or

$$a_2(\alpha) = (1/33)(55 - 5\alpha). \tag{8.45}$$

So, $\overline{A} = \frac{1}{33}(45/50/55)$ is a triangular fuzzy number. To predict 2001 we evaluate $\overline{F}(36)$ obtaining

$$(334.1/354.5, 384.5/405). \tag{8.46}$$

What we have just done is a very small introduction to fuzzy regression, or the fitting of fuzzy functions to fuzzy data. The calculations become more complicated if we had more than two data points.

The fuzzy quadratic function fuzzifies $y = f(x) = ax^2 + bx + c$. The fully fuzzified quadratic will be

$$\overline{Y} = \overline{F}(\overline{X}) = \overline{A} \cdot \overline{X}^2 + \overline{B} \cdot \overline{X} + \overline{C}, \tag{8.47}$$

which we looked at finding its zeros in Chapter 5. We note that $\overline{Y}_e$ will probably not equal $\overline{Y}_i$ (see Example 8.3.3).

Example 8.4.2

A company determines its demand function, for a certain product it produces and sells, to be approximately

$$p = 120 - 0.04x, \tag{8.48}$$

where x is the number of units it can sell per week at price p. Demand functions are never known exactly so we model demand as the fuzzy function

$$\overline{P} = \overline{F}(x) = \overline{A} - \overline{B}x, \tag{8.49}$$

for $\overline{A} = (110/120/130)$ and $\overline{B} = (0.03/0.04/0.05)$ and $x \geq 0$ so that $\overline{P} \geq 0$. That is, x is in some interval $[0, M]$ which guarantees $\overline{P} \geq 0$.

The cost of producing each unit is approximately $C = 10x + 1000$, where x is the number of units produced per week. The variable cost of production (labor, materials, etc.) is around \$10/unit, and the fixed costs (insurance, taxes, utilities, etc.) are about \$1000 per week. These values are always hard to determine, especially the fixed costs, so we model costs as

$$\overline{C} = \overline{G}(x) = \overline{D}x + \overline{E}, \tag{8.50}$$

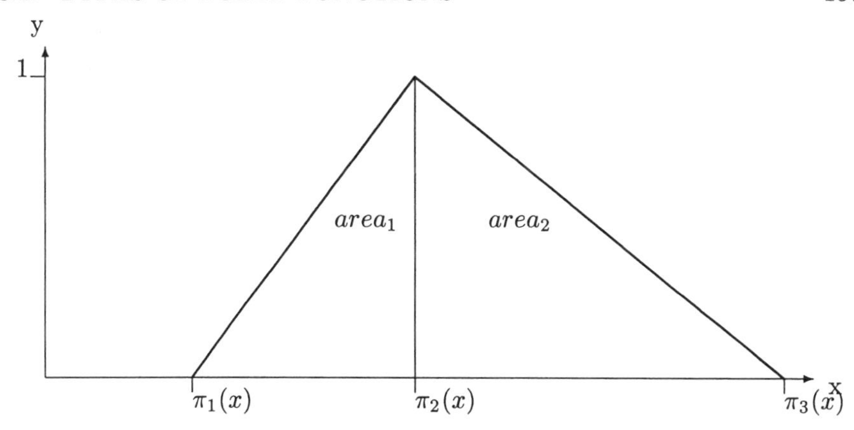

Figure 8.1: The Areas Needed in Example 8.4.2

for $\overline{D} = (9/10/11)$ and $\overline{E} = (900/1000/1100)$. The revenue function is $\overline{P}x$, where now x is the numbers of units we produce and sell each week. We see that the fuzzy profit function is

$$\overline{\Pi} = (\overline{A} - \overline{B}x)x - (\overline{D}x + \overline{E}). \tag{8.51}$$

We wish to find x to maximize $\overline{\Pi}$. However, we can not maximize a fuzzy set. $\overline{\Pi}$ will be a triangular fuzzy number, for each $x \in [0, M]$. Let $\overline{\Pi} = (\pi_1(x)/\pi_2(x)/\pi_3(x))$, for $x \in [0, M]$.

What we do, similar to what is done in finance, where $\overline{\Pi}$ is the probability density of a random variable, is to find x to: (1) $\max(\pi_2(x))$, or maximize the central value of fuzzy profit; (2) maximize the area in the triangle to the right of π_2, or maximize the possibility of obtaining values more than the central value; and (3) minimize the area in the triangle to the left of $\pi_2(x)$, or minimize the possibility of getting results less than the central value. Let $area_1(x) = (\pi_2(x) - \pi_1(x))/2$ and $area_2(x) = (\pi_3(x) - \pi_2(x))/2$. These areas are shown in Figure 8.1. So the optimization problem becomes $\min(area_1(x))$, $\max(\pi_2(x))$ and $\max(area_2(x))$.

We now change $\min(area_1(x))$ to $\max[N - area_1(x)]$, for some large positive number N. The two problems are equivalent. Then $\max(\overline{\Pi})$ becomes

$$\max\{N - area_1(x), \pi_2(x), area_2(x)\}, \tag{8.52}$$

subject to x in some interval $[0, M]$. What is usually done now is for the decision maker to choose values for $\lambda_i > 0$, $i = 1, 2, 3$, and $\lambda_1 + \lambda_2 + \lambda_3 = 1$ and solve the single objective problem

$$\max(\lambda_1(M - area_1(x)) + \lambda_2\pi_2(x) + \lambda_3 area_2(x)), \tag{8.53}$$

for $x \in [0, M]$. The values of the λ_i show for the decision maker the importance of the three goals: $\min(area_1(x))$, $\max(\pi_2)$, and $\max(area_2)$.

This example will be continued in Chapter 16 on fuzzy optimization.

The general fully fuzzified polynomial is

$$\overline{Y} = \overline{F}(\overline{X}) = \overline{A}_n \overline{X}^n + \cdots + \overline{A}_1 \overline{X} + \overline{A}_0, \tag{8.54}$$

for continuous fuzzy numbers $\overline{A}_n, \cdots, \overline{A}_0, \overline{X}$. We expect $\overline{Y}_e \leq \overline{Y}_i$ and $\overline{Y}_e \neq \overline{Y}_i$.

Fuzzy rational functions are quotients of fuzzy polynomials. These fuzzy functions have been previously considered in Examples 8.2.4, 8.2.5 and 8.3.2.

Other special fuzzy functions may be discussed like fuzzifying $f(x) = |x|$, $\sqrt{x}$, etc. Since $y = \sqrt{x}$, $x \geq 0$, is monotone increasing we easily find that $\sqrt{\overline{X}}[\alpha] = [\sqrt{x_1(\alpha)}, \sqrt{x_2(\alpha)}]$ where $\overline{X}[\alpha] = [x_1(\alpha), x_2(\alpha)]$.

The fuzzy exponential is also easy to compute because $y = e^x$ is monotonically increasing. Consider

$$\overline{Y} = \overline{F}(\overline{X}) = \overline{A} \exp(\overline{B} \cdot \overline{X} + \overline{C}), \tag{8.55}$$

for $\overline{A} > 0$, $\overline{X} > 0$ but $\overline{B} < 0$. Then α-cuts of $\overline{Y}_e$ are

$$y_1(\alpha) = a_1(\alpha) \exp(b_1(\alpha) x_2(\alpha) + c_1(\alpha)), \tag{8.56}$$

and

$$y_2(\alpha) = a_2(\alpha) \exp(b_2(\alpha) x_1(\alpha) + c_2(\alpha)). \tag{8.57}$$

We see that $\overline{Y}_e = \overline{Y}_i$ in this example.

Example 8.4.3

The compound interest formula is

$$A = A_0(1+r)^n, \tag{8.58}$$

where A_0 is the initial investment, r is the interest rate (as a decimal) per interest period, n is the number of interest periods and A is the final amount after n interest periods. For this formula the interest rate has to be predicted over n interest periods and it is assumed to be constant over the n period horizon. Let us now model r as a continuous fuzzy number $\overline{r}$ in the interval $(0, 1)$. We will also let A_0 be a continuous fuzzy number $\overline{A}_0$ but we will keep n as a crisp positive integer. Then we have

$$\overline{A} = \overline{A}_0(1+\overline{r})^n. \tag{8.59}$$

Let us show for $\overline{r}$ and $\overline{A}_0$ being fuzzy that we do get this equation. That is, we now derive equation (8.59).

After the first interest period we surely get

$$\overline{A}_1 = \overline{A}_0(1+\overline{r}). \tag{8.60}$$

So, after two interest periods we have

$$\overline{A}_2 = \overline{A}_1 + \overline{A}_1\overline{r}, \tag{8.61}$$

or

$$\overline{A}_2 = \overline{A}_0(1+\overline{r}) + \overline{A}_0(1+\overline{r})\overline{r}, \tag{8.62}$$

which equals

$$\overline{A}_2 = \overline{A}_0(1+\overline{r})[1+\overline{r}], \tag{8.63}$$

or

$$\overline{A}_2 = \overline{A}_0(1+\overline{r})^2, \tag{8.64}$$

since for positive fuzzy numbers multiplication distributes over addition (Section 4.3.4). We therefore see that equation (8.59) holds in general. Equation (8.59) is a fuzzy exponential of the form $\overline{Y} = \overline{A}_0(\overline{a})^x$ for $\overline{A}_0, \overline{a}$ fuzzy but x crisp ($\overline{a} = 1 + \overline{r}$ is the fuzzy base of the fuzzy exponential).

The fuzzy logarithmic function is

$$\overline{Y} = \overline{F}(\overline{X}) = \overline{A} \cdot \ln(\overline{B} \cdot \overline{X} + \overline{C}), \tag{8.65}$$

is also easy to evaluate since $y = \ln(x)$, $x > 0$, is monotonically increasing. To simplify things assume that $\overline{B} > 0$, $\overline{X} > 0$,$\overline{C} > 0$ so that $\overline{B} \cdot \overline{X} + \overline{C} \geq 1$ ($\ln(\overline{B} \cdot \overline{X} + \overline{C}) \geq 0$). Then α-cuts of $\overline{Y}$ are

$$y_i(\alpha) = a_i(\alpha)\ln(b_i(\alpha)x_i(\alpha) + c_i(\alpha)), \tag{8.66}$$

for $i = 1, 2$. In this case we also have $\overline{Y}_e = \overline{Y}_i$.

We now question if the well-known properties of logarithms hold for fuzzy logarithms? If $\overline{A} > 0, \overline{B} > 0$ and n is a positive integer, then are the following equations true or false?

$$\ln(\overline{A} \cdot \overline{B}) = \ln(\overline{A}) + \ln(\overline{B}). \tag{8.67}$$

$$\ln(\overline{A}/\overline{B}) = \ln(\overline{A}) - \ln(\overline{B}). \tag{8.68}$$

$$\ln(\overline{A}^n) = n(\ln(\overline{A})). \tag{8.69}$$

Let us show that equation (8.67) is true. We will use the extension principle extension of $\ln(ab)$, $a > 0$ and $b > 0$. If $\overline{Y}_1 = \ln(\overline{A} \cdot \overline{B})$, $\overline{A}[\alpha] = [a_1(\alpha), a_2(\alpha)]$ and $\overline{B}[\alpha] = [b_1(\alpha), b_2(\alpha)]$ we get the α-cuts of $\overline{Y}_1[\alpha] = [y_{11}(\alpha), y_{12}(\alpha)]$ as

$$y_{11}(\alpha) = \ln(a_1(\alpha)b_1(\alpha)), \tag{8.70}$$

$$y_{12}(\alpha) = \ln(a_2(\alpha)b_2(\alpha)). \tag{8.71}$$

Now set $\overline{Y}_2 = \ln(\overline{A}) + \ln(\overline{B})$, $\overline{Y}_2[\alpha] = [y_{21}(\alpha), y_{22}(\alpha)]$, then

$$y_{21}(\alpha) = \ln(a_1(\alpha)) + \ln(b_1(\alpha)), \tag{8.72}$$

$$y_{22}(\alpha) = \ln(a_2(\alpha)) + \ln(b_2(\alpha)). \tag{8.73}$$

We see that $y_{11}(\alpha) = y_{21}(\alpha)$ and $y_{12}(\alpha) = y_{22}(\alpha)$ for all α. Equation (8.67) is therefore true.

8.4.1 Exercises

1. Show that equation (8.46) in Example 8.4.1 is correct.

2. In Example 8.4.2 let $\overline{P} = (p_1/p_2/p_3)$. Find the domain of $\overline{\Pi}$ in equation (8.51). That is, find M so that $\overline{P} \geq 0$ ($p_1 \geq 0$) if and only if $x \in [0, M]$.

3. In Example 8.4.2 show that $\overline{\Pi}_e = \overline{\Pi}_i$. Use $\overline{A} > 0, \overline{B} > 0, \overline{D} > 0, \overline{E} > 0$ and x in $[0, M]$ of Problem 2.

4. Show that $\overline{Y}_e = \overline{Y}_i$ in equation (8.55).

5. In Example 8.4.3 :

   a. show $\overline{A}_e = \overline{A}_i$ ($\overline{A}_0 > 0, \overline{r} > 0$);

   b. Evaluate equation (8.64) for $\overline{r} = (0.05/0.06, 0.07/0.08)$, $\overline{A}_0 = (90/100/110)$ and $n = 12$ (monthly for one year).

6. In Example 8.4.3 let $\overline{A}_0 = (100/101/102)$ and $\overline{A} = (1980/2000/2020)$ and n being any positive integer. Solve for $\overline{r}$.

   a. Find the classical solution (Chapter 5), $\overline{r}_c$, if it exists.

   b. Find the extension principle solution (Chapter 5) $\overline{r}_e$.

   c. If $\overline{r}_c$ exists, then show $\overline{r}_c \leq \overline{r}_e$.

7. The function $y = f(x) = a_0 \exp(kt)$, for $a_0 > 0, k > 0, t \geq 0$, is called the exponential growth function. Fuzzify it to $\overline{Y} = \overline{F}(t) = \overline{A}_0 \exp(\overline{K}t)$, for $\overline{A}_0 > 0$, $\overline{K} > 0$, $t \geq 0$. We wish to use this fuzzy function to model a fuzzy growth process given by $\overline{F}(0) = \overline{Y}_0$, $\overline{F}(t_1) = \overline{Y}_1$ for $t_1 > 0$. Given this data find $\overline{A}_0, \overline{K}$ if $\overline{Y}_0 = (80/100/120)$, $\overline{Y}_1 = (400/500/600)$.

   a. Use the classical solution method.

   b. Use the extension principle solution.

   c. Compare the two solutions.

8. Same as Problem 7 but now $k < 0$ so it is called an exponential decay function. Given $\overline{F}(0) = \overline{Y}_0 = (90/100/110)$ and $\overline{F}(t_1) = \overline{Y}_1 = (30/40/50)$, then solve for $\overline{A}_0 > 0$ and $\overline{K} < 0$.

   a. Use the classical solution method.

   b. Use the extension principle solution.

   c. Compare the results.

9. Let $\overline{B} = (-12/-10/-8)$, and assume that $\overline{X} < 0, \overline{A} > 0, \overline{C} > 1$. Evaluate $\overline{Y} = \overline{A}\exp(\overline{B} \cdot \overline{X} + \overline{C})$ giving $\overline{Y}_e$ and $\overline{Y}_i$ (the α-cut and interval arithmetic procedure). Do we get $\overline{Y}_e = \overline{Y}_i$?

10. Does $\ln(\overline{A}/\overline{B}) = \ln(\overline{A}) - \ln(\overline{B})$ hold for $\overline{A} > 0, \overline{B} > 0$?

11. Does $\ln(\overline{A}^n) = n\ln(\overline{A})$ hold for $\overline{A} > 0$ and n a positive integer?

12. Consider the fuzzy exponential function $\overline{Y} = \exp(\overline{K}t)$. Determine if the following equations are true or false. $\overline{K}_i$ is a triangular fuzzy number, $i = 1, 2$, and $t \geq 0$.

    a. $(\exp(\overline{K}_1 t))(\exp(\overline{K}_2 t)) = \exp((\overline{K}_1 + \overline{K}_2)t)$.

    b. $\frac{\exp(\overline{K}_1 t)}{\exp(\overline{K}_2 t)} = \exp((\overline{K}_1 - \overline{K}_2)t)$.

    c. $(\exp(\overline{K}_1 t))^n = \exp(n\overline{K}_1 t)$.

    d. $\exp(-\overline{K}_1 t) = \frac{1}{\exp(\overline{K}_1 t)}$.

13. In Example 8.4.1 given any two triangular fuzzy numbers $\overline{Y}_0$ and $\overline{Y}_1$ can we always solve for $\overline{A}$ and $\overline{B}$?

14. Let $\overline{Y} = \exp(\overline{X})$. Does it follow that $\overline{X} = \ln(\overline{Y})$?

15. Let $\overline{Y} = \ln(\overline{X})$, for $\overline{X} > 0$. Will we get $\overline{X} = \exp(\overline{Y})$?

16. Discuss Example 8.4.2 if we allow x, the number of units we produce and sell each week, to be a positive trapezoidal fuzzy number.

17. Discuss Example 8.4.3 if we allow n, the number of interest periods, to be fuzzy.

18. Show for equation (8.65), under the assumptions given in the text, that $\overline{Y}_e = \overline{Y}_i$.

19. In Example 8.4.1 consider using a fuzzy quadratic

$$\overline{Y} = \overline{F}(x) = \overline{A}x^2 + \overline{B}x + \overline{C},$$

    for triangular shaped fuzzy numbers $\overline{A}, \overline{B}, \overline{C}$. Given $\overline{Y} = \overline{Y}_0 > 0$ for $x = 0$, $\overline{Y} = \overline{Y}_1 > 0$ for $x = x_1 > 0$ and $\overline{Y} = \overline{Y}_2 > 0$ for $x = x_2 > x_1$ can we always solve for $\overline{A}$, $\overline{B}$ and $\overline{C}$ so that $\overline{F}(x_i) = \overline{Y}_i$, $i = 0, 1, 2$?

20. Consider the fuzzy function $\overline{Y} = \overline{F}(t) = \overline{A}\ln(\overline{K}t+1)$. Given $\overline{Y} = \overline{Y}_0 > 0$ at $t = 0$ and $\overline{Y} = \overline{Y}_1 > 0$ at $t = t_1 > 0$ solve for $\overline{A}$ and $\overline{K}$ so that $\overline{F}(x_i) = \overline{Y}_i$ for $i = 0, 1$.

    a. Use the classical solution method.

    b. Use the extension principle solution.

    c. Compare the results.

8.5 Inverse Functions

In this section all our fuzzy sets will be fuzzy subsets of the real numbers and we will use only continuous fuzzy numbers as input to fuzzy functions.

Let $\overline{Y} = \overline{F}(\overline{X})$ and $\overline{Y} = \overline{G}(\overline{X})$. We say $\overline{F}$ and $\overline{G}$ are inverses of each other, written $\overline{G} = \overline{F}^{-1}$, if

$$\overline{F}(\overline{G}(\overline{X})) = \overline{X}, \tag{8.74}$$

and

$$\overline{G}(\overline{F}(\overline{X})) = \overline{X}, \tag{8.75}$$

for all $\overline{X}$ for which these equations are defined.

Example 8.5.1

We show that $\overline{Y} = \overline{F}(\overline{X}) = \exp(\overline{X})$ and $\overline{Y} = \overline{F}(\overline{X}) = \ln(\overline{X})$ are inverses of each other. We first argue that

$$\ln(\exp(\overline{X})) = \overline{X}, \tag{8.76}$$

for all continuous fuzzy numbers $\overline{X}$. The extension principle, and the α-cuts and interval arithmetic procedure, both produce the same result here so let use the α-cut method. If $\overline{X}[\alpha] = [x_1(\alpha), x_2(\alpha)]$, then equation (8.76) becomes

$$\ln(\exp([x_1(\alpha), x_2(\alpha)])) = \tag{8.77}$$

$$\ln([\exp(x_1(\alpha)), \exp(x_2(\alpha))]) = \tag{8.78}$$

$$[x_1(\alpha), x_2(\alpha)], \tag{8.79}$$

which is $\overline{X}[\alpha]$, $0 \leq \alpha \leq 1$.

Next we show that for $\overline{X} > 0$

$$\exp(\ln(\overline{X})) = \overline{X}. \tag{8.80}$$

Substituting α-cuts we get

$$\exp(\ln([x_1(\alpha), x_2(\alpha)])) = \tag{8.81}$$

$$\exp([\ln(x_1(\alpha)), \ln(x_2(\alpha))]) = \tag{8.82}$$

$$[x_1(\alpha), x_2(\alpha)], \tag{8.83}$$

an α-cut of $\overline{X}$.

Example 8.5.2

If $y = f(x) = x/(x+1)$, for $x \neq -1$, and if $y = g(x) = x/(1-x)$, for $x \neq 1$, then f and g are inverses of each other.

Let $\overline{Y} = \overline{F}(\overline{X}) = \overline{X}/(\overline{X}+1)$, for -1 not in $\overline{X}[0]$, and let $\overline{Y} = \overline{G}(\overline{X}) = \overline{X}/(1-\overline{X})$, for 1 not in $\overline{X}[0]$. Will $\overline{F}$ and $\overline{G}$ be inverses of each other?

We first use α-cuts and interval arithmetic to evaluate $\overline{F}(\overline{G}(\overline{X}))$ and $\overline{G}(\overline{F}(\overline{X}))$. $\overline{F}(\overline{G}(\overline{X}))$ is

$$\frac{\overline{X}/(1-\overline{X})}{(\overline{X}/(1-\overline{X}))+1}. \tag{8.84}$$

Substitute $[x_1(\alpha), x_2(\alpha)]$ for $\overline{X}[\alpha]$, and first evaluate equation (8.84) assume $0 < \overline{X} < 1$ so that $1-\overline{X} > 0$. Then equation (8.84) becomes

$$[\frac{x_1(\alpha)(1-x_2(\alpha))}{1-x_1(\alpha)}, \frac{x_2(\alpha)(1-x_1(\alpha))}{1-x_2(\alpha)}], \tag{8.85}$$

which is not $\overline{X}[\alpha]$. Therefore $\overline{F}$ and $\overline{G}$ are not inverses of each other using the α-cut method.

Next try the extension principle to evaluate $\overline{F}(\overline{G}(\overline{X}))$. This is the extension of $h(x) = f(g(x))$ which is

$$h(x) = \frac{x/(1-x)}{(x/(1-x))+1}. \tag{8.86}$$

Assume $x \neq 1$, and 1 does not belong to $\overline{X}[0]$, then h is a continuous function and let $\overline{Y} = \overline{H}(\overline{X})$, the extension of h. We find α-cuts of $\overline{Y}$ as

$$y_1(\alpha) = \min\{h(x) = x | x \in \overline{X}[\alpha]\}, \tag{8.87}$$

$$y_2(\alpha) = \max\{h(x) = x | x \in \overline{X}[\alpha]\}. \tag{8.88}$$

We now see that $y_i(\alpha) = x_i(\alpha)$ for all α and $i = 1, 2$. Hence, $\overline{F}(\overline{G}(\overline{X})) = \overline{X}$. Also, we may compute, using the extension principle, $\overline{G}(\overline{F}(\overline{X})) = \overline{X}$, and conclude $\overline{G} = \overline{F}^{-1}$.

Example 8.5.3

Let $y = f(x) = ax + b$ and $y = g(x) = (x-b)/a$, for $a \neq 0$. The functions f and g are inverses of each other. Fuzzify to the completely fuzzified $\overline{Y} = \overline{F}(\overline{X}) = \overline{A} \cdot \overline{X} + \overline{B}$ and $\overline{Y} = \overline{G}(\overline{X}) = (\overline{X} - \overline{B})/\overline{A}$, for zero not in $\overline{A}[0]$. Are $\overline{F}$ and $\overline{G}$ inverses of each other?

Using the α-cut procedure to compute $\overline{F}(\overline{G}(\overline{X}))$ we obtain

$$\overline{A}(\frac{\overline{X}-\overline{B}}{\overline{A}})+\overline{B}, \tag{8.89}$$

which can not equal $\overline{X}$ since $\overline{A}/\overline{A}\neq 1$ and $-\overline{B}+\overline{B}\neq 0$.

Try the extension principle method on $\overline{F}(\overline{G}(\overline{X}))$ which is the extension of

$$h(x)=a(\frac{x-b}{a})+b, \tag{8.90}$$

for $a\neq 0$. Since h is a continuous function, the α-cuts of $\overline{Y}=\overline{F}(\overline{G}(\overline{X}))$ are

$$y_1(\alpha)=\min\{h(x)=x|x\in\overline{X}[\alpha],a\in\overline{A}[\alpha],b\in\overline{B}[\alpha]\}, \tag{8.91}$$

$$y_2(\alpha)=\max\{h(x)=x|x\in\overline{X}[\alpha],a\in\overline{A}[\alpha],b\in\overline{B}[\alpha]\}. \tag{8.92}$$

Hence $y_i(\alpha)=x_i(\alpha)$ for all α and $i=1,2$. This means that $\overline{F}(\overline{G}(\overline{X}))=\overline{X}$. In a similar manner we can show $\overline{G}(\overline{F}(\overline{X}))=\overline{X}$. By the extension principle procedure $\overline{G}=\overline{F}^{-1}$.

8.5.1 Exercises

1. Let $\overline{Y} = \overline{F}(\overline{X}) = (\overline{X})^2$ for $\overline{X} \geq 0$ and $\overline{Y} = \overline{G}(\overline{X}) = \sqrt{\overline{X}}$ for $\overline{X} \geq 0$. Show that $\overline{F}(\overline{G}(\overline{X})) = \overline{X}$ and $\overline{G}(\overline{F}(\overline{X})) = \overline{X}$ for $\overline{X} \geq 0$.

   a. Using the extension principle.

   b. Using the α-cut method.

2. Show that equation (8.85) is correct when $0 < \overline{X} < 1$. Find the correct expression if $\overline{X} > 1$ and for $\overline{X} < 0$. In either case do we get $\overline{F}(\overline{G}(\overline{X})) = \overline{X}$?

3. Using the extension principle also compute $\overline{G}(\overline{F}(\overline{X}))$ in Example 8.5.2.

4. Show that equations (8.76) and (8.80) are also correct using the extension principle.

5. Let $y = f(x) = (2x+1)/(3-x)$, for $x \neq 3$, and let $y = g(x) = (3x-1)/(2+x)$, for $x \neq -2$. f and g are inverses of each other. Extend f to $\overline{F}$ and g to $\overline{G}$. Compute $\overline{F}(\overline{G}(\overline{X}))$ and $\overline{G}(\overline{F}(\overline{X}))$:

   a. using the extension principle, and

   b. using the α-cut method.

   c. Does either procedure give $\overline{G} = \overline{F}^{-1}$?

6. Let $y = f(x) = \sqrt{ax-b}$ for $a > 0, b > 0, x \geq b/a$, and let $y = g(x) = (x^2+b)/a$. f and g are inverses of each other. Fuzzify to $\overline{Y} = \overline{F}(\overline{X}) = \sqrt{\overline{A} \cdot \overline{X} - \overline{B}}$ for $\overline{A} > 0, \overline{B} > 0, \overline{A} \cdot \overline{X} - \overline{B} \geq 0$, and $\overline{Y} = \overline{G}(\overline{X}) = (\overline{X}^2 + \overline{B})/\overline{A}$ for $\overline{X} \geq 0$. Determine if we get $\overline{G} = \overline{F}^{-1}$:

   a. using the extension principle, and

   b. using the α-cut and interval arithmetic method.

7. Let $y = f(x) = x^2 + 6x - 10$ for $x \geq -3$ and $y = g(x) = -3 + \sqrt{x+19}$ for $x \geq -19$. f and g are inverses of each other. Extend f to $\overline{F}$ and extend g to $\overline{G}$. Is it true that $\overline{G} = \overline{F}^{-1}$?

   a. Using the extension principle?

   b. Using the α-cut method?

8. Let $g = f^{-1}$. Give a description of f and g so that we will obtain $\overline{X} = \overline{F}(\overline{G}(\overline{X})) = \overline{G}(\overline{F}(\overline{X}))$ using α-cuts and interval arithmetic.

9. Let $g = f^{-1}$. Will the extension principle always give $\overline{G} = \overline{F}^{-1}$?

10. Let $y = f(x) = x^3$ and $y = g(x) = x^{\frac{1}{3}}$. f and g are inverses of each other. Extend f to $\overline{F}$ and g to $\overline{G}$. Find $\overline{F}(\overline{G}(\overline{X})))$ and $\overline{G}(\overline{F}(\overline{X}))$:

a. using the extension principle, and

b. using the α-cut method.

c. Does either procedure give $\overline{G} = \overline{F}^{-1}$?

11. Let $y = f(t) = ae^{kt}$ for $a > 0$, $k > 0$ and $y = g(t) = (1/k)\ln(t/a)$ for $t > 0$.

    a. Extend f to $\overline{F}$ and g to $\overline{G}$. Use both methods of extension. Do we get that $\overline{G} = \overline{F}^{-1}$?

    b. Do the same as in part a but extend f to $\overline{F}(\overline{T}) = \overline{A}\exp[\overline{K}\,\overline{T}]$ and g to $\overline{G}(\overline{T}) = \overline{K}^{-1}\ln[\overline{T}/\overline{A}]$.

8.6 Derivatives

Throughout this section $\overline{X} = \overline{F}(t)$ for t a real number in some interval D in $\mathbf{R}$, $\overline{F}$ a fuzzy function and $\overline{X}$ a continuous fuzzy number. For example, $\overline{F}(t) = \overline{A}t + \overline{B}$, or $\overline{F}(t) = \overline{A}\exp(\overline{B}t)$, or $\overline{F}(t) = \overline{A}t^2 + \overline{B}t$, for continuous fuzzy numbers $\overline{A}$ and $\overline{B}$. In this section all fuzzy expressions will be evaluated using α-cuts and interval arithmetic.

We first discuss possible definitions of $d\overline{F}/dt$, the derivative of fuzzy function $\overline{F}$ with respect to t. Let us start with the method used in calculus. If $y = f(x)$, then df/dx at x is defined to be

$$\lim_{h \to 0} \frac{f(x+h) - f(x)}{h}, \tag{8.93}$$

provided the limit exists.

Try this for $\overline{X} = \overline{F}(t)$. First compute $(\overline{F}(t+h) - \overline{F}(t))/h$ and then take the limit as h approaches zero. This method is not going to work because even in the simplest case of $\overline{X} = \overline{F}(t) = \overline{A}t + \overline{B}$ there is no limit. Suppose that $\overline{A} > 0$, $t \geq 0$ and also $h > 0$. Then

$$(\overline{F}(t+h) - \overline{F}(t))/h, \tag{8.94}$$

equals

$$[(\overline{A} - \overline{A})t + (\overline{B} - \overline{B})]/h + \overline{A}. \tag{8.95}$$

We know that $(\overline{A} - \overline{A})t \neq 0$, unless $t = 0$, and $\overline{B} - \overline{B} \neq 0$ assuming $\overline{A}$ and $\overline{B}$ are fuzzy (their support is not a single point). The expression in equation (8.95) has no limit as h approaches zero through the positive numbers. This definition of $d\overline{F}/dt$ does not exist.

As an alternate procedure of defining $d\overline{F}/dt$ let $\overline{F}(t)[\alpha] = [x_1(t,\alpha), x_2(t,\alpha)]$ be the α-cuts, which are now considered to be functions of both t and α. For example, let $\overline{X} = \overline{F}(t) = \overline{A}\exp(-\overline{K}t)$ for $\overline{A}[\alpha] = [a_1(\alpha), a_2(\alpha)]$ and $\overline{K}[\alpha] = [k_1(\alpha), k_2(\alpha)]$. Then $\overline{F}(t)[\alpha] = [a_1(\alpha)\exp(-k_2(\alpha)t), a_2(\alpha)\exp(-k_1(\alpha)t)]$ where $x_1(t,\alpha) = a_1(\alpha)\exp(-k_2(\alpha)t)$ and $x_2(t,\alpha) = a_2\exp(-k_1(\alpha)t)$. Then we define

$$\frac{d\overline{F}}{dt}[\alpha] = [\partial x_1/\partial t, \partial x_2/\partial t], \tag{8.96}$$

provided this interval defines the α-cuts of a continuous fuzzy number for all $t \in D$. Then the derivative of $\overline{F}$ at t is a continuous fuzzy number whose α-cuts are given by the intervals

$$[\partial x_1(t,\alpha)/\partial t, \partial x_2(t,\alpha)/\partial t]. \tag{8.97}$$

For equation (8.97) to define the α-cuts of a continuous fuzzy number we must have: (1) $\partial x_1/\partial t$ is a continuous monotonically increasing function of α, $0 \leq \alpha \leq 1$, for each t in D; (2) $\partial x_2/\partial t$ is a continuous monotonically decreasing function of α, $0 \leq \alpha \leq 1$, for each t in D; and (3)

$\partial x_1(t,1)/\partial t \leq \partial x_2(t,1)/\partial t$ for all t in D. In other words: (1) $\partial^2 x_1/\partial\alpha\partial t > 0$; (2) $\partial^2 x_2/\partial\alpha\partial t < 0$; and (3) same as (3) above.

Example 8.6.1

Let $\overline{X} = \overline{F}(t) = \overline{A}t + \overline{B}$, $t \geq 0$ and for $\overline{A} = (-10/-8,-7/-5)$ and $\overline{B} = (16/20,22/26)$. We see that $\overline{A}[\alpha] = [-10+2\alpha, -5-2\alpha]$ and $\overline{B}[\alpha] = [16+4\alpha, 26-4\alpha]$. Then the α-cuts of $\overline{F}(t)$ are

$$[(16+4\alpha)+(-10+2\alpha)t, (26-4\alpha)+(-5-2\alpha)t], \tag{8.98}$$

so that

$$\frac{d\overline{F}}{dt}[\alpha] = [-10+2\alpha, -5-2\alpha], \tag{8.99}$$

which is $\overline{A}[\alpha]$. Hence the derivative exists and $d\overline{F}/dt = \overline{A}$ for $t \geq 0$.

Example 8.6.2

Let $\overline{X} = \overline{F}(t) = \overline{A}t^2 + \overline{B}t + \overline{C}$ for $t \geq 0$. Also let $\overline{A}[\alpha] = [a_1(\alpha), a_2(\alpha)]$, $\overline{B}[\alpha] = [b_1(\alpha), b_2(\alpha)]$ and $\overline{C}[\alpha] = [c_1(\alpha), c_2(\alpha)]$. Then $\overline{F}(t)[\alpha] = [L(\alpha), R(\alpha)]$ where

$$L(\alpha) = a_1(\alpha)t^2 + b_1(\alpha)t + c_1(\alpha), \tag{8.100}$$

$$R(\alpha) = a_2(\alpha)t^2 + b_2(\alpha)t + c_2(\alpha). \tag{8.101}$$

Then

$$\frac{d\overline{F}}{dt}[\alpha] = [2a_1(\alpha)t + b_1(\alpha), 2a_2(\alpha)t + b_2(\alpha)]. \tag{8.102}$$

Hence

$$\frac{d\overline{F}}{dt} = 2\overline{A}t + \overline{B}, \tag{8.103}$$

for $t \geq 0$. If $t < 0$, then we may get different results.

Example 8.6.3

Let $\overline{X} = \overline{F}(t) = (\overline{A}t)/(t+\overline{B})$ for $\overline{A} = (10/20/30)$ and $\overline{B} = (6/7/8)$ and assume that $t \geq 0$. Then α-cuts of $\overline{X}$ are

$$[\frac{a_1(\alpha)t}{t+b_2(\alpha)}, \frac{a_2(\alpha)t}{t+b_1(\alpha)}]. \tag{8.104}$$

Next we find $\partial x_1/\partial t$ and $\partial x_2/\partial t$ giving the interval

$$[\frac{a_1(\alpha)b_2(\alpha)}{(t+b_2(\alpha)^2)}, \frac{a_2(\alpha)b_1(\alpha)}{(t+b_1(\alpha))^2}]. \tag{8.105}$$

We calculate $\overline{B}[\alpha] = [6+\alpha, 8-\alpha]$ and $\overline{A}[\alpha] = [10+10\alpha, 30-10\alpha]$. Substitute these values into equation (8.105) and then compute $\partial^2 x_1/\partial\alpha\partial t$ which turns out to be positive, and then find $\partial^2 x_2/\partial\alpha\partial t$ which is negative. Let $\alpha = 1$ in equation (8.105) and it follows that $\partial x_1(t,1)/\partial t = \partial x_2(t,1)/\partial t$. This means that equation (8.105) defines the α-cuts of a continuous fuzzy number $d\overline{F}/dt$ for all $t \geq 0$.

Example 8.6.4

Let $\overline{X} = \overline{F}(t) = \overline{A}/\sqrt{t}$ for $t > 0$. Then

$$\overline{F}(t)[\alpha] = [\frac{a_1(\alpha)}{\sqrt{t}}, \frac{a_2(\alpha)}{\sqrt{t}}], \tag{8.106}$$

so that $[\partial x_1/\partial t, \partial x_2/\partial t]$ equals

$$[\frac{-a_1(\alpha)}{2t^{3/2}}, \frac{-a_2(\alpha)}{2t^{3/2}}]. \tag{8.107}$$

But, the function on the left side of the interval in equation (8.107) is a decreasing function of α. Therefore, $d\overline{F}/dt$ does not exist.

Example 8.6.5

Let $\overline{X} = \overline{F}(t) = \overline{A}\exp(-\overline{K}t)$ for $\overline{A} > 0$, $\overline{K} > 0$ and $t \geq 0$. Writing $\overline{A}[\alpha] = [a_1(\alpha), a_2(\alpha)]$ and $\overline{K}[\alpha] = [k_1(\alpha), k_2(\alpha)]$ we obtain the α-cuts

$$\overline{F}(t)[\alpha] = [a_1(\alpha)\exp(-k_2(\alpha)t), a_2(\alpha)\exp(-k_1(\alpha)t)], \tag{8.108}$$

so that we can compute $[\partial x_1/\partial t, \partial x_2/\partial t]$ and get

$$[-a_1(\alpha)k_2(\alpha)\exp(-k_2(\alpha)t), -a_2(\alpha)k_1(\alpha)\exp(-k_1(\alpha)t)]. \tag{8.109}$$

Now let $\overline{A} = (90/100/110)$ and $\overline{K} = (\frac{1}{3}/\frac{1}{2}/\frac{2}{3})$. Then $\overline{A}[\alpha] = [90 + 10\alpha, 110 - 10\alpha]$ and $\overline{K}[\alpha] = [\frac{1}{3} + \frac{1}{6}\alpha, \frac{2}{3} - \frac{1}{6}\alpha]$. Substitute these results into equation (8.109) and compute $\partial^2 x_1/\partial\alpha\partial t$ and we see that it is positive for

$t \in [0, T)$ for some $T > 0$. Next we determine $\partial^2 x_2/\partial\alpha\partial t$ and it is negative for $t \in [0, T)$. The derivative of $\overline{F}$ with respect to t exists for t in $[0, T)$.

Example 8.6.6

Let $\overline{X} = \overline{F}(t) = \overline{A}\cdot\ln(\overline{B}t+\overline{C})$ for $\overline{A} > 0, \overline{B} > 0, \overline{C} > 0, t > 0$ and $\overline{B}t+\overline{C} > 1$. Then if $\overline{F}(t)[\alpha] = [L(\alpha), R(\alpha)]$, then

$$L(\alpha) = a_1(\alpha)\ln(b_1(\alpha)t + c_1(\alpha)), \tag{8.110}$$

$$R(\alpha) = a_2(\alpha)\ln(b_2(\alpha)t + c_2(\alpha)), \tag{8.111}$$

so that the interval $[\partial x_1/\partial t, \partial x_2/\partial t]$ is

$$[\frac{a_1(\alpha)b_1(\alpha)}{b_1(\alpha)t + c_1(\alpha)}, \frac{a_2(\alpha)b_2(\alpha)}{b_2(\alpha)t + c_2(\alpha)}]. \tag{8.112}$$

Now assume that $\overline{A} = (70/100/130)$, $\overline{B} = (20/40/60)$ and $\overline{C} = (4/5/6)$. Find the α-cuts of $\overline{A}$, $\overline{B}$ and $\overline{C}$, then substitute into equation (8.112). Then compute $\partial^2 x_1/\partial\alpha\partial t$ and $\partial^2 x_2/\partial\alpha\partial t$. The first one is positive and the second is negative for all $t > 0$. We also check to see that the two end points in equation (8.112) are equal for $\alpha = 1$. $d\overline{F}/dt$ exists for $t > 0$.

In the rest of this section let us turn to solving elementary differential equations having fuzzy initial conditions. Consider solving

$$\frac{dx}{dt} = kx, \tag{8.113}$$

for $x(0) = a$. The unique solution for $x = x(t)$ is

$$x = ae^{kt}. \tag{8.114}$$

Now assume the initial condition is fuzzy so that $x(0) = \overline{A}$ a continuous fuzzy number. Solve for $\overline{X} = \overline{F}(t)$.

As in Chapter 5 we can consider three solution methods: (1) classical solution $\overline{X}_c$, (2) the extension principle solution $\overline{X}_e$, and (3) the α-cut and interval arithmetic solution $\overline{X}_i$. We only consider the first two in this section. We illustrate these two methods through the following example. Assume $\overline{A} = (a_1/a_2/a_3)$.

Example 8.6.7

Solve equation (8.113) subject to $x(0) = \overline{A}$, $\overline{A}[\alpha] = [a_1(\alpha), a_2(\alpha)]$, first using the classical procedure. Let $\overline{X}(t) = [x_1(t,\alpha), x_2(t,\alpha)]$ and substitute into equation (8.113). Assume that now $k > 0$. We get

$$[\partial x_1/\partial t, \partial x_2/\partial t] = [kx_1, kx_2], \tag{8.115}$$

or

$$\frac{\partial x_i}{\partial t} = kx_i, \tag{8.116}$$

for $i = 1, 2$. Solving for the x_i we get

$$x_i(t,\alpha) = a_i e^{kt}, \tag{8.117}$$

for $i = 1, 2$. The initial conditions are

$$x_1(0,\alpha) = a_1(\alpha), \tag{8.118}$$

$$x_2(0,\alpha) = a_2(\alpha). \tag{8.119}$$

Hence

$$x_i(t,\alpha) = a_i(\alpha)e^{kt}, \tag{8.120}$$

for $i = 1, 2$.

If the intervals

$$[a_1(\alpha)e^{kt}, a_2(\alpha)e^{kt}], \tag{8.121}$$

define a continuous fuzzy number for all t in some interval D, we say the classical solution $\overline{X}_c(t)$ exists and its $\alpha - cuts$ are given by equation (8.121). These intervals do define a continuous fuzzy numbers so we conclude that

$$\overline{X}_c(t) = \overline{A}e^{kt}. \tag{8.122}$$

Now let us use the extension principle method. This procedure fuzzifies, using the extension principle, the crisp solution. The unique crisp solution is $x = x(t) = ae^{kt}$ where $x(0) = a$. Hence $\overline{X}_e = \overline{A}e^{kt}$. We will now demand that $\overline{X}_e$ must solve the original differential equation to be called a "solution". That is, we say the extension principle solution exists when $\overline{X}_e(t) = \overline{A}e^{kt}$ satisfies the differential equation and fuzzy initial conditions. We use α-cuts and interval arithmetic to check to see if $\overline{X}_e(t)$ satisfies equation (8.113). Now $\overline{X}(t)[\alpha] = [a_1(\alpha)e^{kt}, a_2(\alpha)e^{kt}]$ so that

$$\frac{\partial}{\partial t}[a_1(\alpha)e^{kt}, a_2(\alpha)e^{kt}] = \tag{8.123}$$

$$k[a_1(\alpha)e^{kt}, a_2(\alpha)e^{kt}], \tag{8.124}$$

is true and we see $\overline{X}(0) = \overline{A}$. Hence, $\overline{X}_e(t)$ exists as a solution and also $\overline{X}_c = \overline{X}_e$.

8.6.1 Exercises

1. In Example 8.6.1 assume $t < 0$. Does $d\overline{F}/dt$ exist for $t < 0$?

2. In Example 8.6.2 assume $t < 0$. Show now that $d\overline{F}/dt$ may not exist for all $t < 0$. Let $\overline{A} = (1/2/3)$ and $\overline{B} = (50/100/150)$.

3. In Example 8.6.3 check to see if $d\overline{F}/dt$ exists for $t < 0$.

4. Let $\overline{X} = \overline{F}(t) = \overline{A}\sqrt{t}$, $t \geq 0$ and $\overline{A} \approx (a_1/a_2, a_3/a_4)$. Find $d\overline{F}/dt$.

5. In Example 8.6.5 find T.

6. Let $\overline{X} = \overline{F}(t) = \overline{A}\exp(\overline{K}t)$ for $\overline{A} > 0$, $\overline{K} > 0$ and $t \geq 0$. Show that $d\overline{F}/dt$ exists for all $t \geq 0$ and find its α-cuts.

7. Rework Example 8.6.6 for $\overline{A} = (2/3/4)$, $\overline{B} = (5/6/7)$ and $\overline{C} = (8/9/10)$.

8. In Example 8.6.3, find $d\overline{F}/dt$ at $t = 1$.

9. In Example 8.6.5, find $d\overline{F}/dt$ at $t = 0$.

10. In Example 8.6.6, find $d\overline{F}/dt$ at $t = 1$.

11. Work Example 8.6.7 if $k < 0$. Does $\overline{X}_c(t)$ exist? Does $\overline{X}_e(t)$ satisfy the differential equation?

12. Consider

$$\frac{dx}{dt} = kx^2,$$

for $k > 0$, $t \geq 0$ and $x(0) = a > 0$. The unique solution is

$$x = a/(1 - akt).$$

Now set $x(0) = \overline{A} > 0$ for the fuzzy initial condition.

  a. Does $\overline{X}_c(t)$ exist? If so, find its α-cuts.
  b. Find $\overline{X}_e(t)$, find its α-cuts and determine if it satisfies the differential equation.

13. Consider

$$\frac{dx}{dt} = k(x - a),$$

for $k > 0$, $x > a$, $t \geq 0$ and $x(0) = c > a$. The unique solution is

$$x = a + (c - a)e^{kt}.$$

Now let $x(0) = \overline{C} > a$, a fuzzy initial condition.

a. Does $\overline{X}_c(t)$ exist? If so, find its α-cuts.

b. Find $\overline{X}_e(t)$, find its α-cuts and determine if it satisfies the differential equation.

Chapter 9

Fuzzy Plane Geometry

Crisp plane geometry starts with points, then lines and parallel lines, circles, triangles, rectangles, etc. In fuzzy plane geometry we will do the same. Our fuzzy points, lines, circles, etc. will all be fuzzy subsets of $\mathbf{R} \times \mathbf{R}$. We assume the standard $xy-$ rectangular coordinate system in the plane. Since fuzzy subsets of $\mathbf{R} \times \mathbf{R}$ will be surfaces in $\mathbf{R}^3$ we can not easily present graphs of their membership functions. However, α-cuts of fuzzy subsets of $\mathbf{R} \times \mathbf{R}$ will be crisp subsets of the plane. Using an xy-coordinate system we may draw pictures of α-cuts of fuzzy points in $\mathbf{R} \times \mathbf{R}$, also for fuzzy lines, etc. In this way we can see what the membership functions of fuzzy subsets of $\mathbf{R} \times \mathbf{R}$ look like.

A fuzzy point $\overline{P}$ at (x_0, y_0) will generalize the idea of a triangular shaped fuzzy number. $\overline{P}$ is defined by its membership function, written $z = \overline{P}(x, y)$, for (x, y) in $\mathbf{R}^2$ and $z \in [0, 1]$. The constraints on the membership function are:

1. $z = \overline{P}(x, y)$ is continuous,
2. $\overline{P}(x, y) = 1$ if and only if $(x, y) = (x_0, y_0)$, and
3. $\overline{P}[\alpha]$ is closed, bounded and convex, $0 \leq \alpha \leq 1$.

The second condition means $\overline{P}$ is a fuzzy point at (x_0, y_0). In the third condition: (1) closed means the set contains its boundary; (2) bounded means there is a positive M so that $\overline{P}[\alpha]$ is a subset of a circle, center at $(0, 0)$, of radius M; and (3) convex means given any two points in $\overline{P}[\alpha]$, the straight line connecting these points also lies entirely in $\overline{P}[\alpha]$. A circle together with its inside is closed, bounded and convex. A circle, with a hole in it, is not convex. Figure 9.1 gives other examples of convex, and not convex, subsets of the plane.

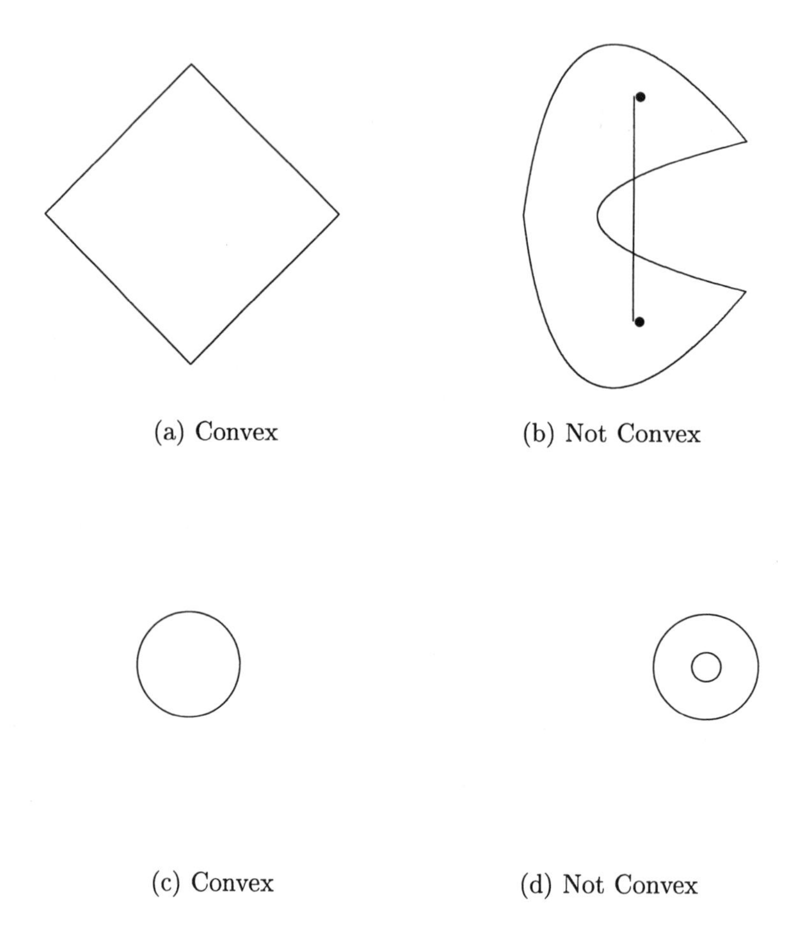

Figure 9.1: Examples of Convex Sets

Example 9.1

Let $\overline{M}$ and $\overline{N}$ be two triangular shaped fuzzy numbers $\overline{M} \approx (m_1/m_2/m_3), \overline{N} \approx (n_1/n_2/n_3)$. Then

$$\overline{P}(x,y) = \min(\overline{M}(x), \overline{N}(y)), \tag{9.1}$$

is a fuzzy point at (m_2, n_2).

Example 9.2

Define $\overline{P}$ by

$$\overline{P}(x,y) = \begin{cases} 1 - x^2 - y^2, & \text{if } x^2 + y^2 \leq 1, \\ 0, & \text{otherwise.} \end{cases} \tag{9.2}$$

$\overline{P}$ is a fuzzy point at $(0,0)$. The graph of $z = \overline{P}(x,y)$ is a right circular cone with vertex at $(0,0)$, with all points inside and on the circle $x^2 + y^2 = 1$.

A fuzzy line $\overline{L}$ will generalize the equation $ax + by = c$ whose graph is a straight line as long as a or b are not both zero. We use the extension principle to define $z = \overline{L}(x,y)$. Let $\overline{A}$, $\overline{B}$ and $\overline{C}$ be triangular shaped fuzzy numbers, then

$$\overline{L}(x,y) = \max\{\min\{\overline{A}(a), \overline{B}(b), \overline{C}(c)\} | ax + by = c\}. \tag{9.3}$$

α-cuts of $\overline{L}$ may be found from

$$\overline{L}[\alpha] = \{(x,y) | ax + by = c, a \in \overline{A}[\alpha], b \in \overline{B}[\alpha], c \in \overline{C}[\alpha]\}, \tag{9.4}$$

$0 \leq \alpha \leq 1$. $\overline{L}[1]$ will be a crisp line, and $\overline{L}[0]$ will be base of the fuzzy line.

Example 9.3

A "fat" fuzzy line. Let $\overline{A} = (-1/0/1)$, $\overline{B} = (-1/1/2)$ and $\overline{C} = (0/1/2)$. Then $\overline{L}[0] = \mathbf{R^2}$ and $\overline{L}[1]$ is the line $0x + y = 1$, or $y = 1$.

Example 9.4

A "thin" fuzzy line. Let $\overline{A} = 2$, $\overline{B} = 1$ and $\overline{C} = (0/1/2)$. $\overline{A}$ and $\overline{B}$ are crisp (real) numbers. Then $\overline{L}[1]$ is the line $2x + y = 1$. $\overline{L}[0]$ is the infinite strip in the xy-plane bordered by $y = -2x$ and $y = -2x + 2$.

We say fuzzy line $\overline{L}$ contains fuzzy point $\overline{P}$ if $\overline{P} \leq \overline{L}$ ($\overline{P}(x,y) \leq \overline{L}(x,y)$, for all (x,y)). Suppose $\overline{P}$ is a fuzzy point at (x_0,y_0) and $\overline{P} \leq \overline{L}$. Then we may easily show (x_0,y_0) belongs to $\overline{L}[1]$ because

$$1 = \overline{P}(x_0,y_0) \leq \overline{L}(x_0,y_0). \tag{9.5}$$

We know when two crisp lines are parallel. Fuzzy lines may be parallel to varying degrees. Let $\overline{L_a}$ and $\overline{L_b}$ be two fuzzy lines. A measure of the parallelness (ρ) of $\overline{L}_a$ and $\overline{L}_b$ is defined to be $1-\lambda$ where

$$\lambda = \sup_{x,y}\{\min(\overline{L_a}(x,y), \overline{L_b}(x,y))\}. \tag{9.6}$$

We see that λ is just the height of the intersection (using T_m) of $\overline{L_a}$ and $\overline{L_b}$. If $\overline{L_a} \cap \overline{L_b} = \phi$, then $\rho = 1$ and $\overline{L_a}$ and $\overline{L_b}$ are completely parallel. Suppose l_a (l_b) is the crisp line in $\overline{L_a}[1](\overline{L_b}[1])$. If l_a and l_b intersect, then $\rho = 0$ and $\overline{L_a}, \overline{L_b}$ are definitely not parallel.

Example 9.5

Let $\overline{A} = -2$, $\overline{B} = 1$, $\overline{C_1} = (6/8/10)$, and $\overline{C_2} = (8/10/12)$. Let $\overline{L_i}$ be $-2x + y = \overline{C_i}$, $i = 1,2$. Then we calculate $\rho = 0.5$

For fuzzy circles we fuzzify $(x-a)^2 + (y-b)^2 = d^2$ which is the equation of a circle, center at (a,b), with radius $d \geq 0$. A fuzzy circle $\overline{Cir}$ is defined by its membership function

$$\overline{Cir}(x,y) = \max\{\min\{\overline{A}(a), \overline{B}(b), \overline{D}(d)\}|(x-a)^2 + (y-b)^2 = d^2\}, \tag{9.7}$$

for continuous fuzzy numbers $\overline{A}$, $\overline{B}$, $\overline{D} \geq 0$. The fuzzy circle has center at $(\overline{A}, \overline{B})$ and radius $\overline{D}$. α-cuts of $\overline{Cir}$ may be found from

$$\overline{Cir}[\alpha] = \{(x,y)|(x-a)^2 + (y-b)^2 = d^2, a \in \overline{A}[\alpha], b \in \overline{B}[\alpha], d \in \overline{D}[\alpha]\}, \tag{9.8}$$

$0 \leq \alpha \leq 1$. $\overline{Cir}[1]$ is a crisp circle and $\overline{Cir}[0]$ is the base of the fuzzy circle.

Example 9.6

A "fat" fuzzy circle. Let $\overline{A} = \overline{B} = \overline{D} = (0/1/2)$. $\overline{Cir}[1]$ is the crisp circle $(x-1)^2 + (y-1)^2 = 1$. $\overline{Cir}[0]$, the base of the fuzzy circle, is a region, inside the rectangle $[-2,4] \times [-2,4]$, with no holes in it. As we increase α from zero to one will be obtain $\overline{Cir}[\alpha]$ with holes in them. The limit of the $\overline{Cir}[\alpha]$, as α approaches 1, is the crisp circle with the hole, $(x-1)^2 + (y-1)^2 < 1$, inside it.

Example 9.7

A "regular" fuzzy circle. By regular we mean a fuzzy circle $\overline{Cir}$ whose α-cuts, $0 \leq \alpha \leq 1$, all have holes in them. Let $\overline{A} = \overline{B} = 1$ (crisp one) and $\overline{D} = (1/2/3)$. Then $\overline{Cir}[0] = \{(x,y)|1 \leq (x-1)^2 + (y-1)^2 \leq 9\}$. Similarly, $\overline{Cir}[\alpha] = \{(x,y)|(1+\alpha)^2 \leq (x-1)^2 + (y-1)^2 \leq (3-\alpha)^2\}$, $0 \leq \alpha \leq 1$.

We may define the fuzzy area of a fuzzy circle. Let $\overline{area}$ be a fuzzy subset of $\mathbf{R}$ defined by its membership function

$$\overline{area}(z) = \max\{\min(\overline{A}(a), \overline{B}(b), \overline{D}(d))|z = \pi d^2, d = radius\ of\ circle\ at\ (a,b)\}, \tag{9.9}$$

whose α-cuts can be formed by

$$\overline{area}[\alpha] = \{\pi d^2 | d = radius\ of\ circle\ at\ (a,b), a \in \overline{A}[\alpha], b \in \overline{B}[\alpha], d \in \overline{D}[\alpha]\}, \tag{9.10}$$

$0 \leq \alpha \leq 1$.

To study fuzzy triangles and fuzzy rectangles we need the idea of a fuzzy line segment $\overline{L}_{PQ}$ from fuzzy point $\overline{P}$ to fuzzy point $\overline{Q}$. We will not define the membership function for $\overline{L}_{PQ}$ directly but instead we define the α-cuts of $\overline{L}_{PQ}$. Define

$$\overline{L}_{PQ}[\alpha] = \{l | l\ is\ a\ line\ segment\ from\ a\ point\ in \overline{P}[\alpha]\ to\ a\ point\ in \overline{Q}[\alpha]\}, \tag{9.11}$$

$0 \leq \alpha \leq 1$. If $\overline{P}[0] \subset \overline{Q}[0]$, or even if $\overline{P}[0] \cap \overline{Q}[0] \neq \phi$, we may not get a "regular" fuzzy line segment. So, let us assume $\overline{P}[0]$ and $\overline{Q}[0]$ are disjoint in the definition of an α-cut of $\overline{L}_{PQ}$.

Now we can give the definition of a fuzzy triangle in terms of its α-cuts. Let $\overline{P}, \overline{Q}, \overline{R}$ be three fuzzy points so that $\overline{P}[0], \overline{Q}[0], \overline{R}[0]$ are mutually disjoint. Let $\overline{L}_{PQ}, \overline{L}_{QR}, \overline{L}_{RP}$ be fuzzy line segments. $\overline{P}, \overline{Q}, \overline{R}$ define a fuzzy triangle $\overline{T}$ whose α-cuts are

$$\overline{T}[\alpha] = \overline{L}_{PQ}[\alpha] \cup \overline{L}_{QR}[\alpha] \cup \overline{L}_{PR}[\alpha], \tag{9.12}$$

or generally

$$\overline{T} = \overline{L}_{PQ} \cup \overline{L}_{QR} \cup \overline{L}_{PR}, \tag{9.13}$$

where we use max, for t-conorm, for union.

Example 9.10

$\overline{P}, \overline{Q}, \overline{R}$ will be fuzzy points at $v_1 = (0,0), v_2 = (0,1), v_3 = (1,0)$, respectively. Each fuzzy point is like the fuzzy point in Example 9.2. $\overline{P}(\overline{Q}, \overline{R})$ has a membership function whose graph is a surface, which is a right circular cone, radius of the base circle is 0.1 and vertex is at $v_1(v_2, v_3)$. Then $\overline{T}$ is the fuzzy triangle generated by $\overline{P}, \overline{Q}, \overline{R}$. $\overline{T}[0]$ is a triangular subset of $\mathbf{R}^2$ with a triangular hole in it. $\overline{T}[1]$ is the crisp triangle with vertices $(0,0), (1,0), (0,1)$.

9.1 Exercises

1. In Example 9.1 let $\overline{M} = (-4/1/2)$ and $\overline{N} = (6/7/9)$. Find $\overline{P}[0.5]$ and $\overline{P}[1]$. By "find" we mean draw pictures of these α-cuts with respect to an xy-coordinate system.

2. In Example 9.2, find $\overline{P}[0], \overline{P}[0.5], \overline{P}[1]$. By "find" we mean draw a picture with respect to an xy-coordinate system.

3. Show that α-cuts of $\overline{L}$ defined in equation (9.3) are given by equation (9.4).

4. In Example 9.3 show that $\overline{L}[0]$ is the whole xy-plane.

5. In Example 9.4 draw a picture of $\overline{L}[0]$, $\overline{L}[0.5]$ and $\overline{L}[1]$.

6. In this problem draw pictures of $\overline{L}[0], \overline{L}[0.5]$ and $\overline{L}[1]$. $\overline{A} = (-1/0/1)$, $\overline{B} = (-1/1/2)$ and $\overline{C} = (0/1/2)$. An equation like $\overline{A}x + \overline{B}y = \overline{C}$ is defined to be a fuzzy line $\overline{L}$ as in equation (9.3). Find α-cuts from equation (9.4).

   a. $\overline{A}x + y = 1$.
   b. $2x + \overline{B}y = 1$.
   c. $\overline{A}x + \overline{B}y = 1$.
   d. $\overline{A}x + y = \overline{C}$.
   e. $2x + \overline{B}y = \overline{C}$.

7. Is the fuzzy point $\overline{P}$ in Example 9.2 contained in any of the fuzzy lines given in problem 6?

8. There can be alternate definitions of a fuzzy line. For the following specifications of a line give a definition of the corresponding fuzzy line (equation (9.3)) and its α-cuts (equation (9.4)):

   a. $y = mx + b$;
   b. $y - v = m(x - u)$, point (u, v), slope $=m$; and
   c. $\frac{y-v_1}{x-u_1} = \frac{u_2-v_1}{u_2-u_1}$, two points (u_1, v_1), (u_2, v_2).

9. In Example 9.5 show that $\rho = 0.5$.

10. In $2x + \overline{B}y = 1$, let $\overline{L_1}$ have $\overline{B_1} = (-1/1/2)$ and $\overline{L_2}$ have $\overline{B_2} = (-2/-1/1)$. Find λ in equation (9.6) and hence ρ the measure of parallelness of $\overline{L_1}$ and $\overline{L_2}$.

11. In this problem draw picture of $\overline{Cir}[0]$, $\overline{Cir}[0.5]$ and $\overline{Cir}[1]$. Let $\overline{A} = \overline{B} = \overline{D} = (0/1/2)$. An equation like $(x - \overline{A})^2 + (y - \overline{B})^2 = \overline{D}$ is to be defined as a fuzzy circle as in equation (9.7) with α-cuts in equation (9.8).

a. $(x - \overline{A})^2 + (y - 1)^2 = 1.$

b. $(x - 1)^2 + (y - \overline{B})^2 = 1$.

c. $(x - \overline{A})^2 + (y - \overline{B})^2 = 1.$

d. $(x - \overline{A})^2 + (y - \overline{B})^2 = \overline{C}^2.$

e. $(x - 1)^2 + (y - \overline{B})^2 = \overline{C}^2.$

12. Let $\overline{perimeter}$ be a fuzzy subset of $\mathbf{R}$ which is to be the fuzzy circumference of a fuzzy circle $\overline{Cir}$. Give a definition for fuzzy circumference, as in equation (9.9) for fuzzy area, and give a way to find α-cuts of fuzzy circumference, as equation (9.10) for fuzzy area.

13. Give a definition of the fuzzy area of a fuzzy triangle and then an expression to find its α-cuts.

14. Give a definition of a fuzzy rectangle based on the definition of a fuzzy triangle.

15. Draw a picture of $\overline{T}[0]$ in Example 9.10.

16. Draw pictures of $\overline{T}[0], \overline{T}[0.5]$ and $\overline{T}[1]$ if:

    a. $\overline{P}$ as in Example 9.10, $\overline{Q} = (0, 1), \overline{R} = (1, 0)$ crisp points; and

    b. $\overline{P}$ as in Example 9.10, $\overline{Q} = (0, 1), \overline{R}$ as in Example 9.10.

17. Let $\mathcal{T}$ be all crisp triangles in regular plane geometry. Define R, a crisp relation on $\mathcal{T}$, as $R(t_1, t_2) = 1$ if and only if t_1 and t_2 are similar, for $t_1, t_2 \in \mathcal{T}$. Now let $\overline{\mathcal{T}}$ be all fuzzy triangles. Define the concept of "similar" between fuzzy triangles and then define an equivalence relation R on $\overline{\mathcal{T}}$ as $R(\overline{t_1}, \overline{t_2}) = 1$ if and only if $\overline{t_1}$ and $\overline{t_2}$ are similar.

18. In crisp trigonometry we can define sine, cosine and tangent of an angle θ in terms of the ratios of the sides of a right triangle. For example, $\sin(\theta)$, one of the angles in the right triangle not equal to the 90 degree angle, would be the ratio of the opposite side to the hypotenuse. Define the idea of a fuzzy right triangle. The side lengths now become continuous fuzzy numbers. Let $\overline{\theta}$ be one of the fuzzy angles in this fuzzy right triangle, not the fuzzy 90 degree angle. We can then define fuzzy sine as $\sin(\overline{\theta})$ being the ratio of the fuzzy length of the opposite side to the fuzzy hypotenuse. Continue this way to define fuzzy cosine, fuzzy tangent and develop fuzzy trigonometry from this point.

19. All our definitions in the text used t-norm $T_m = \min$. Consider using another t-norm: T_b, T_p or T^*. Which one, if any, will minimize the fuzziness (also needs to be defined) of a fuzzy line, a fuzzy circle, etc.?

20. Consider the following alternate definition of a fuzzy circle. Let $\overline{A}$, $\overline{B}$ and $\overline{D}$ be continuous fuzzy numbers. A fuzzy circle $\overline{Cir}$ is the set of all ordered pairs $(\overline{X}, \overline{Y})$, where $\overline{X}$ and $\overline{Y}$ are both continuous fuzzy numbers, which satisfy

$$(\overline{X} - \overline{A})^2 + (\overline{Y} - \overline{B})^2 = \overline{D}^2.$$

Is this a "good", or "not so good", definition of a fuzzy circle?

Chapter 10

Fuzzy Trigonometry

10.1 Introduction

This chapter completes our study of the basic fuzzy functions. In Section 8.4 we had the fuzzy functions from linear to logarithmic. The next section introduces the fuzzy trigonometric functions, their inverses (see Section 8.5) and their derivatives (see Section 8.6). We also see to what extent the basic crisp trigonometric identities will hold for the fuzzy trigonometric functions. Finally, the fuzzy hyperbolic trigonometric functions are in the third section.

10.2 Standard Fuzzy Trigonometry

We first want to fuzzify $y = \sin(x)$ and then $y = \cos(x)$ and $y = \tan(x)$. We will always use radians (real numbers) for x since that is what is needed in calculus. So let $\overline{X}$ be a continuous fuzzy number and set $\overline{Y} = \overline{F}(\overline{X}) = \sin(\overline{X})$. We will only use the extension principle to find $\overline{Y}$ because the α-cut and interval arithmetic method (Section 8.3) requires an algorithm, using a finite number of arithmetic operations, to approximate $y = \sin(x)$ to a certain accuracy. This algorithm can involve using a power series for $y = \sin(x)$. Also we would need to employ similar algorithms for the other trigonometric functions if we wanted to find $\overline{Y}$ from the α-cut procedure. We will omit the α-cut method from this section but you are asked to try it for $y = \sin(x)$ in the exercises.

If $\overline{Y} = [y_1(\alpha), y_2(\alpha)]$, then from the extension principle we may find the α-cuts of $\overline{Y} = \sin(\overline{X})$ as follows

$$y_1(\alpha) = \min\{\sin(x) | x \in \overline{X}[\alpha]\}, \tag{10.1}$$

and

$$y_2(\alpha) = \max\{\sin(x) | x \in \overline{X}[\alpha]\}, \tag{10.2}$$

for $0 \leq \alpha \leq 1$. Since $y = \sin(x)$ is periodic with period 2π we can get very different results for $\overline{Y}$ depending on the size of the support and core of $\overline{X}$.

Example 10.2.1

Let $\overline{X} = (0/\frac{\pi}{2}, \frac{3\pi}{2}/2\pi)$. Then we calculate that

$$\overline{Y}(x) = \begin{cases} 1, & -1 \leq x \leq 1, \\ 0, & \text{otherwise.} \end{cases} \tag{10.3}$$

$\overline{Y}$ is not a continuous fuzzy number. $\overline{Y}$ is a fuzzy interval.

Example 10.2.2

Now let $\overline{X} = (0/\frac{\pi}{2}/\pi)$. Then we find that

$$\overline{Y}[\alpha] = [\sin(\frac{\pi}{2}\alpha), 1], \tag{10.4}$$

for $0 \leq \alpha \leq 1$, and $\overline{Y}(x) = 0$ if $x \leq 0$ or $x > 1$. Again $\overline{Y}$ is not a continuous fuzzy number.

Example 10.2.3

Set $\overline{X} = (0/\frac{\pi}{4}/\frac{\pi}{2})$. Then we see that

$$\overline{Y}[\alpha] = [\sin(\frac{\pi}{4}\alpha), \sin(\frac{\pi}{2} - \frac{\pi}{4}\alpha)], \tag{10.5}$$

for $0 \leq \alpha \leq 1$, and $\overline{Y}(x) = 0$ if $x \leq 0$ or $x \geq 1$. $\overline{Y}$ is a continuous fuzzy number.

In a similar way we may compute $\overline{Y} = \cos(\overline{X})$ for various values of $\overline{X}$. To fuzzify $y = \tan(x)$ let $\overline{X}$ be a triangular (or trapezoidal) fuzzy number in the interval $(-\pi/2, \pi/2)$. Then if $\overline{Y} = \tan(\overline{X})$ we obtain α-cuts as

$$y_1(\alpha) = \min\{\tan(x) | x \in \overline{X}[\alpha]\}, \tag{10.6}$$

and

$$y_2(\alpha) = \max\{\tan(x) | x \in \overline{X}[\alpha]\}, \tag{10.7}$$

for all $\alpha \in [0, 1]$. Since $y = \tan(x)$ is monotonically increasing on $(-\pi/2, \pi/2)$ we get $y_1(\alpha) = \tan(x_1(\alpha))$, and $y_2(\alpha) = \tan(x_2(\alpha))$ where $\overline{X}[\alpha] = [x_1(\alpha), x_2(\alpha)]$.

We can not expect crisp trigonometric identities like $\sin^2(x)+\cos^2(x)=1$ and $\tan^2(x)+1=sec^2(x)$ to hold for fuzzy trigonometric functions. For example, $\sin^2(\overline{X})+\cos^2(\overline{X})\neq 1$ because the left side of this equation is a fuzzy subset of the real numbers and the right side is not fuzzy, it is the crisp number one. But we can expect one to belong to the core of $\sin^2(\overline{X})+\cos^2(\overline{X})$.

Example 10.2.4

If $\overline{X}=(0/\frac{\pi}{4}/\frac{\pi}{2})$, then we found the α-cuts of $\sin(\overline{X})$ in Example 10.2.3. Now let $\overline{Y}=\cos(\overline{X})$. Then we determine that the α-cuts of $\cos(\overline{X})$ are

$$\overline{Y}[\alpha]=[\cos(\frac{\pi}{2}-\frac{\pi}{4}\alpha),\cos(\frac{\pi}{4}\alpha)], \tag{10.8}$$

since $y=\cos(x)$ is monotonically decreasing on $[0,\pi/2]$. Because $\sin(\overline{X})\geq 0$ and $\cos(\overline{X})\geq 0$ we may easily find α-cuts of both $\sin^2(\overline{X})$ and $\cos^2(\overline{X})$ as

$$\sin^2(\overline{X})[\alpha]=[\sin^2(\frac{\pi}{4}\alpha),\sin^2(\frac{\pi}{2}-\frac{\pi}{4}\alpha)], \tag{10.9}$$

and

$$\cos^2(\overline{X})[\alpha]=[\cos^2(\frac{\pi}{2}-\frac{\pi}{4}\alpha),\cos^2(\frac{\pi}{4}\alpha)]. \tag{10.10}$$

We see that $\sin^2(\overline{X})\approx(0/0.5/1)$ and $\cos^2(\overline{X})\approx(0/0.5/1)$. Hence $\sin^2(\overline{X})+\cos^2(\overline{X})\approx(0/1/2)$. The real number one belongs to the core of $\sin^2(\overline{X})+\cos^2(\overline{X})$.

Although the fuzzy trigonometric identities do not hold exactly, the fuzzy trigonometric functions are still periodic

$$\sin(\overline{X}+2\pi)=\sin(\overline{X}), \tag{10.11}$$

$$\cos(\overline{X}+2\pi)=\cos(\overline{X}), \tag{10.12}$$

and

$$\tan(\overline{X}+\pi)=\tan(\overline{X}). \tag{10.13}$$

Now let us look at the derivatives of the fuzzy trigonometric functions using the methods of Section 8.6. Consider $\overline{X}=\overline{F}(t)=\sin(\overline{A}t)$, $t\geq 0$, $\overline{A}>0$ and $\overline{A}$ a trapezoidal fuzzy number. We let the α-cuts of $\sin(\overline{A}t)$ be $[x_1(t,\alpha),x_2(t,\alpha)]$ and then the α-cuts of $d\overline{F}/dt$ are defined to be $[\partial x_1/\partial t,\partial x_2/\partial t]$, provided that these intervals define the α-cuts of a continuous fuzzy number. However, we will end up as in Example 10.2.1 because the core and support of $\overline{A}t$ will grow and exceed a length of 2π. That is, $t(a_2-a_1)$ and $t(a_4-a_3)$, where $\overline{A}=(a_1/a_2,a_3/a_4)$, will get larger and larger having $\sin(\overline{A}t)$ look like what we got, equation (10.3), in Example 10.2.1. Then $x_1(t,\alpha)=-1$ and $x_2(t,\alpha)=1$ for all α and then $\partial x_1/\partial t=\partial x_2/\partial t=0$

all α. $d\overline{F}/dt$ will eventually be identically zero. We surely do not want this to happen.

To control for the size of the core and support of the input to sine we will instead consider $\overline{X} = \overline{F}(t) = \sin(\overline{A}(t))$, where $\overline{A}(t)$ is a function of t, not $\overline{A}$ times t. For example, let $\overline{A}(t) = (t - 0.1/t/t + 0.1)$ for $0.1 \leq t \leq (\frac{\pi}{2} - 0.1)$. Now the support of $\overline{A}(t)$ is always of length 0.2 and the core is always just a single point.

Example 10.2.5

Let $\overline{A}(t) = (t - 0.1/t/t + 0.1)$ for $0.1 \leq t \leq (\frac{\pi}{2} - 0.1)$. As in Example 10.2.3 we get the α-cuts of $\sin(\overline{A}(t))$ as

$$[\sin(t - 0.1 + 0.1\alpha), \sin(t + 0.1 - 0.1\alpha)]. \tag{10.14}$$

So, the α-cuts of $d\overline{F}/dt$ would be

$$[\cos(t - 0.1 + 0.1\alpha), \cos(t + 0.1 - 0.1\alpha)], \tag{10.15}$$

if these intervals define a continuous fuzzy number. But $\cos(t - 0.1 + 0.1\alpha)$ decreases as α increases. Hence $d\overline{F}/dt$ does not exist for this $\overline{A}(t)$.

Example 10.2.6

Use the same $\overline{A}(t)$ and interval for t as in Example 10.2.5. Let $\overline{X} = \overline{F}(t) = \tan(\overline{A}(t))$. The α-cuts of $\overline{X}$ are

$$[\tan(t - 0.1 + 0.1\alpha), \tan(t + 0.1 - 0.1\alpha)], \tag{10.16}$$

whose partials on t give the α-cuts

$$[sec^2(t - 0.1 + 0.1\alpha), sec^2(t + 0.1 - 0.1\alpha)]. \tag{10.17}$$

These intervals do define a continuous fuzzy number. $d\overline{F}/dt$ exists and its α-cuts are given by equation (10.17).

The last thing we look at in this section is the fuzzification of the inverse trigonometric functions. Let $y = f(x) = \sin(x)$ for $-\pi/2 \leq x \leq \pi/2$ and let its inverse function be $y = g(x) = \arcsin(x)$ for $-1 \leq x \leq 1$. Define $\overline{Y} = \overline{F}(\overline{X}) = \sin(\overline{X})$ for $\overline{X}$ a continuous fuzzy number in the interval $[-\pi/2, \pi/2]$ and also define $\overline{Y} = \overline{G}(\overline{X}) = \arcsin(\overline{X})$ where now $\overline{X}$ is a continuous fuzzy number in $[-1, 1]$. Do we obtain $\overline{F}(\overline{G}(\overline{X})) = \overline{X}$ and $\overline{G}(\overline{F}(\overline{X})) = \overline{X}$? The fuzzy function $\overline{Y} = \overline{G}(\overline{F}(\overline{X}))$ is the extension principal extension of $y = g(f(x)) = \arcsin(\sin(x))$. But for x in $[-\pi/2, \pi/2]$ we know $g(f(x)) = x$ so that $\overline{G}(\overline{F}(\overline{X})) = \overline{X}$. Similarly, we see that $\overline{F}(\overline{G}(\overline{X})) = \overline{X}$. Hence we have $\overline{G}(\overline{X}) = \overline{F}^{-1}(\overline{X})$.

Example 10.2.7

Let $\overline{X}[\alpha] = [x_1(\alpha), x_2(\alpha)]$ the α-cut of a continuous fuzzy number in $[-\pi/2, \pi/2]$. Since $y = \sin(x)$ is increasing on the interval $[-\pi/2, \pi/2]$ we obtain

$$\sin(\overline{X})[\alpha] = [\sin(x_1(\alpha)), \sin(x_2(\alpha))]. \tag{10.18}$$

Then the α-cuts of $\arcsin(\sin(\overline{X}))$ are

$$\arcsin[\sin(x_1(\alpha)), \sin(x_2(\alpha))], \tag{10.19}$$

which is, since $\sin(x_1(\alpha)) \leq sin(x_2(\alpha))$

$$[\arcsin(\sin(x_1(\alpha)), \arcsin(\sin(x_2(\alpha))], \tag{10.20}$$

or

$$[x_1(\alpha), x_2(\alpha)], \tag{10.21}$$

an α-cut of $\overline{X}$.

10.2.1 Exercises

1. Let $I = [-0.1, 0.1]$. Then $y = h(x) = x - x^3/6$ will approximate $y = \sin(x)$ on I. In fact

$$|\sin(x) - h(x)| < (8.3)10^{-8},$$

for x in I. Use $h(x)$, α-cuts and interval arithmetic, to calculate $\sin(\overline{X})$ for $\overline{X}$ a triangular fuzzy number in I. Compare your result to the extension principle method. Take three cases: (1) $\overline{X} > 0$; (2) $\overline{X} < 0$; and (3) zero belongs to the support of $\overline{X}$.

2. Show that $\sin(\overline{X})$ in Example 10.2.1 produces the $\overline{Y}$ given in equation (10.3).

3. Show that $\sin(\overline{X})$ in Example 10.2.2 has the α-cuts given by equation (10.4).

4. Show that the α-cuts of $\sin(\overline{X})$ in Example 10.2.3 are those in equation (10.5).

5. Let $\overline{A}$ be a triangular fuzzy number in $(0, \pi/2)$. Define $s = \tan(\overline{A})[1]$ and $t = sec(\overline{A})[1]$, the ($\alpha = 1$)-cut. Show that $s^2 + 1 = t^2$.

6. Show that equations (10.11)–(10.13) are correct.

7. Show that the intervals in equation (10.17) do define a continuous fuzzy number.

8. Find $\overline{Y} = \cos(\overline{X})$ for:

   a. $\overline{X} = (0/\frac{\pi}{2}, \frac{3\pi}{2}/2\pi)$,

   b. $\overline{X} = (0/\frac{\pi}{2}/\pi)$, and

   c. $\overline{X} = (0/\frac{\pi}{4}/\frac{\pi}{2})$.

9. Let $\overline{A}(t) = (t - 0.1/t/t + 0.1)$ for $t \in [0.1, \pi/2 - 0.1]$. Define $\overline{Y} = \overline{F}(t) = \cos(\overline{A}(t))$. Find $d\overline{F}/dt$ or show that it does not exist.

10. Use the same $\overline{A}(t)$ as in Problem 9. Define $\overline{Y} = \overline{F}(t) = sec(\overline{A}(t))$, Find $d\overline{F}/dt$ or show that it does not exist.

11. In Example 10.2.5 let $\overline{A} = (t - 0.1/t/t + 0.1)$ but for t in the interval $[-\pi/2 + 0.1, -0.1]$. Does $d\overline{F}/dt$ exist?

12. In Example 10.2.6 use the same $\overline{A}(t)$ but the interval $[-\pi/2 + 0.1, -0.1]$. Does $d\overline{F}/dt$ exist?

13. Let $y = f(x) = \tan(x)$ for $x \in (-\pi/2, \pi/2)$ and $y = g(x) = \arctan(x)$ for x any real number. Fuzzify f to $\overline{F}$ for $\overline{X}$ a triangular fuzzy number in $(-\pi/2, \pi/2)$ and fuzzify g to $\overline{G}$ for $\overline{X}$ a triangular fuzzy subset of the real numbers. Show using the extension principle that $\overline{G} = \overline{F}^{-1}$.

14. Let $\overline{Y} = \overline{F}(t) = \arcsin(\overline{A}(t))$ for $\overline{A}(t)$ a function of t in some interval I so that $\overline{A}(t)$ remains in $[-1, 1]$. Does $d\overline{F}/dt$ exist? Investigate for various functions for $\overline{A}(t)$.

15. Let $\overline{X}$ be a continuous fuzzy number. Is $\sin(\overline{X})$ and $\cos(\overline{X})$ always a fuzzy subset, not necessarily a fuzzy number, of the interval $[-1, 1]$?

16. Let $\overline{X}$ be a continuous fuzzy number. Determine if the following equations are true or false. Justify your answer.

    a. $\sin(-\overline{X}) = -\sin(\overline{X})$.
    b. $\cos(-\overline{X}) = \cos(\overline{X})$.
    c. $\sin(\overline{X} + \pi/2) = \cos(\overline{X})$.

17. In Chapter 9 we introduced fuzzy triangles. All the crisp trigonometric functions can be defined for angles, between zero degrees and ninety degrees, from a right triangle. Define a fuzzy right triangle and fuzzy angles. From this define fuzzy sine, cosine and tangent. Continue in this way to develop fuzzy trigonometric and compare to the fuzzy trigonometric in this section. See Problem 18 in Section 9.1.

18. All the fuzzy trigonometric functions were defined with the extension principle using the t-norm min. Define, and develop, some of their basic properties as in this section, using the extension principle but with another t-norm T_b or T_p or T^*.

19. Show that:

    a. $-1 \leq sin(\overline{X}) \leq 1$, and
    b. $-1 \leq cos(\overline{X}) \leq 1$.

20. Solving fuzzy trigonometric equations. Given the crisp trigonometric equation $2\cos(x) = 1$ we know that one solution for x is $x = \pi/3$. So consider a fuzzy trigonometric equation

$$2\cos(\overline{X}) = \overline{A},$$

for $\overline{A} = (0.5/1/1.5)$ which is approximately one. Solve for $\overline{X}$ a continuous fuzzy number in $[0, \pi/2]$.

    a. Using the classical solution method (Section 5.2.1).
    b. Using the extension principle method (Section 5.2.2).

10.3 Hyperbolic Trigonometric Functions

The basic hyperbolic trigonometric functions, written as $\sinh(x)$, $\cosh(x)$ and $\tanh(x)$, are important in calculus. In this section we fuzzify them, study their inverses and derivatives.

The hyperbolic sine is defined as $y = \sinh(x) = (e^x - e^{-x})/2$. Since $dy/dx > 0$ for all x we see that it is monotonically increasing for all x with $\sinh(0) = 0$. If $\overline{X}$ is a continuous fuzzy number with $\overline{X}[\alpha] = [x_1(\alpha), x_2(\alpha)]$, then it is easy to find α-cuts of $\sinh(\overline{X})$ because they are

$$[\sinh(x_1(\alpha)), \sinh(x_2(\alpha))], \tag{10.22}$$

for $0 \leq \alpha \leq 1$.

The hyperbolic cosine is $y = \cosh(x) = (e^x + e^{-x})/2$. Now $\cosh(x)$ is: (1) monotonically decreasing for $x < 0$, (2) monotonically increasing for $x > 0$, and (3) $\cosh(0) = 1$. So it is easy to find α-cuts of $\cosh(\overline{X})$ if $\overline{X} < 0$ or $\overline{X} > 0$. For example, if $\overline{X} < 0$, then α-cuts of $\cosh(\overline{X})$ are

$$[\cosh(x_2(\alpha)), \cosh(x_1(\alpha))]. \tag{10.23}$$

The hyperbolic tangent is $y = \tanh(x) = \frac{\sinh(x)}{\cosh(x)}$. The function $\tanh(x)$ is: (1) monotonically increasing for all x, (2)$\lim_{x \to -\infty} \tanh(x) = -1$, (3) $\lim_{x \to \infty} \tanh(x) = 1$, and (4) $\tanh(0) = 0$. Like $\sinh(x)$ it is easy to find α-cuts of $\tanh(\overline{X})$ for $\overline{X}$ being a continuous fuzzy number.

Now let $\overline{X} = \overline{F}(t) = \sinh(\overline{A}(t))$, for $\overline{A}(t)$ being a continuous fuzzy number for t a real number in some interval. We wish to see when $d\overline{F}/dt$ exists. We will need the following results: (1) the derivative of $\sinh(x)$ is $\cosh(x)$, (2) the derivative of $\cosh(x)$ is $\sinh(x)$, and (3) the derivative of $\tanh(x)$ is $(\cosh(x))^{-2} = (\mathrm{sech}(x))^2$, the square of the hyperbolic secant.

Let $\overline{A}(t)[\alpha] = [a_1(t, \alpha), a_2(t, \alpha)]$ so that $\overline{F}(t)[\alpha] = [\sinh(a_1(t, \alpha)), \sinh(a_2(t, \alpha))]$. Then $d\overline{F}/dt$ has α-cuts equal to $[L(t, \alpha), R(t, \alpha)]$ where

$$L(t, \alpha) = \cosh(a_1(t, \alpha))\frac{\partial a_1}{\partial t}, \tag{10.24}$$

and

$$R(t, \alpha) = \cosh(a_2(t, \alpha))\frac{\partial a_2}{\partial t}, \tag{10.25}$$

as long as $[L, R]$ defines the α-cuts of a continuous fuzzy number for all t. For $d\overline{F}/dt$ to exist we need $\partial L/\partial\alpha > 0$, $\partial R/\partial\alpha < 0$ for $0 \leq \alpha \leq 1$ and $L(t, 1) = R(t, 1)$ for all t.

Example 10.3.1

Let $\overline{A}(t) = (t^2/t^2 + t/t^2 + 2t)$ for $t \geq 0$. Then $\overline{A}(t)[\alpha] = [t^2 + t\alpha, t^2 + 2t - t\alpha]$. If $\overline{X} = \overline{F}(t) = \sinh(\overline{A}(t))$, then we compute L and R from equations (10.24) and (10.25) to be

$$L(t,\alpha) = \cosh(t^2 + t\alpha)(2t + \alpha), \tag{10.26}$$

and

$$R(t,\alpha) = \cosh(t^2 + 2t - t\alpha)(2t + 2 - \alpha). \tag{10.27}$$

Then we find that $\partial L/\partial\alpha > 0$, $\partial R/\partial\alpha < 0$ for $0 \leq \alpha \leq 1$ and for $t \geq 0$, and also $L(t,1) = R(t,1)$ for all $t \geq 0$. Hence, $d\overline{F}/dt$ exists with α-cuts given by the intervals $[L(t,\alpha), R(t,\alpha)]$.

Two hyperbolic trigonometric identities are $(\cosh(x))^2 - (\sinh(x))^2 = 1$ and $(\tanh(x))^2 + (\operatorname{sech}(x))^2 = 1$. Let us look at how the identity $(\cosh(x))^2 - (\sinh(x))^2 = 1$ holds when we substitute $\overline{X} \geq 0$ a continuous fuzzy number for x. Let $\overline{X}[\alpha] = [x_1(\alpha), x_2(\alpha)]$. Assume that $\overline{X} \approx (x_1/x_2/x_3)$. Then α-cuts of $(\cosh(\overline{X}))^2 - (\sinh(\overline{X}))^2$ are

$$[(\cosh(x_1(\alpha)))^2 - (\sinh(x_2(\alpha)))^2, (\cosh(x_2(\alpha)))^2 - (\sinh(x_1(\alpha)))^2]. \tag{10.28}$$

If $\alpha = 1$, then $x_1(1) = x_2(1) = x_2$ and interval (10.28) becomes a single point

$$(\cosh(x_2))^2 - (\sinh(x_2))^2 = 1. \tag{10.29}$$

The fuzzy hyperbolic trigonometric identity is satisfied at $\alpha = 1$.

We may specify inverses for the hyperbolic trigonometric functions. For example, $\sinh^{-1}(x) = \ln(x+\sqrt{x^2+1})$ for all real numbers x, and $\tanh^{-1}(x) = \frac{1}{2}\ln([1+x]/[1-x])$ for $-1 < x < 1$. The function $y = \tanh^{-1}(x)$ may be used in (fuzzy) neural nets as shown in Chapter 13. So let $y = f(x) = \tanh(x)$ and $y = g(x) = \tanh^{-1}(x)$. Fuzzify both to $\overline{Y} = \overline{F}(\overline{X})$, for $\overline{X}$ a continuous fuzzy number and $\overline{Y} = \overline{G}(\overline{X})$, now $\overline{X}$ a continuous fuzzy number in the interval $(-1,1)$. Using the extension principle we can show that $\overline{G} = \overline{F}^{-1}$.

10.3.1 Exercises

1. Find α-cuts of $\cosh(\overline{X})$ for:

   a. $\overline{X}$ a continuous fuzzy number and $\overline{X} > 0$;

   b. $\overline{X} = (-1/0/2)$.

2. Find α-cuts of $\tanh(\overline{X})$, for $\overline{X}$ a continuous fuzzy number whose α-cuts are $[x_1(\alpha), x_2(\alpha)]$.

3. In Example 10.3.1 show that $L(t,\alpha)$ (and $R(t,\alpha)$) is an increasing (decreasing) function of α.

4. Let $\overline{A}(t) = (\sin(t) - t/\sin(t)/\sin(t) + t)$ for $t \geq 0$. If $\overline{F}(t) = \sinh(\overline{A}(t))$ does $d\overline{F}/dt$ exist for all $t \geq 0$?

5. Let $\overline{A}(t) = (t^2/t^2 + t/t^2 + 2t)$ for $t \geq 0$. If $\overline{F}(t) = \tanh(\overline{A}(t))$, then does $d\overline{F}/dt$ exist for all real numbers $t \geq 0$?

6. Let $\overline{A}(t)$ be the same as in Problem 5. If $\overline{F}(t) = \cosh(\overline{A}(t))$ does $d\overline{F}/dt$ exist for $t \geq 0$.

7. Let $\overline{X} \approx (x_1/x_2/x_3)$ a triangular shaped fuzzy number. Let "a" belong to the ($\alpha = 1$)-cut of $\tanh(\overline{X})$ and let "b" belong to the ($\alpha = 1$)-cut of $\operatorname{sech}(\overline{X})$. Recall that $\operatorname{sech}(x) = (\cosh(x))^{-1}$. Show that $a^2 + b^2 = 1$. This verifies $(\tanh(\overline{X}))^2 + (\operatorname{sech}(\overline{X}))^2 = 1$ at the $\alpha = 1$ level.

8. Show, using the extension principle, that for continuous fuzzy number $\overline{X}$ we get $\sinh(\sinh^{-1}(\overline{X})) = \sinh^{-1}(\sinh(\overline{X})) = \overline{X}$.

9. Does $2\sinh(\ln(\overline{X})) = \overline{X} - (\overline{X})^{-1}$ for continuous fuzzy number $\overline{X} > 0$.

10. Find $\sinh^{-1}(\overline{X})$ by showing its α-cuts. Assume $\overline{X}$ is a continuous fuzzy number.

Chapter 11

Systems of Fuzzy Linear Equations

In this chapter we are interested in solutions to systems of fuzzy linear equations. To keep things as simple as possible we will only work with 2×2 systems, or two equations with two unknowns. As in Chapter 5 we will consider three types of solution: classical, extension principle, and the α-cut and interval arithmetic method. Throughout this chapter $\overline{A} \leq \overline{B}$ means $\overline{A}(x) \leq \overline{B}(x)$ for all x.

We first review the crisp theory for 2×2 systems and then fuzzify it. A 2×2 linear system is written as

$$a_{11}x_1 + a_{12}x_2 = b_1, \tag{11.1}$$

$$a_{21}x_1 + a_{22}x_2 = b_2, \tag{11.2}$$

for crisp constants $a_{11}, \cdots, a_{22}, b_1$ and b_2. We are to solve these two equations simultaneously for x_1 and x_2.

Throughout this chapter we will make the necessary assumptions so that the system of linear equations has a unique solution. The needed assumption here is that $a_{11}a_{22} - a_{21}a_{12} \neq 0$. Then the unique solution may be written

$$x_1 = \frac{b_1 a_{22} - b_2 a_{12}}{a_{11}a_{22} - a_{21}a_{12}}, \tag{11.3}$$

$$x_2 = \frac{b_2 a_{11} - b_1 a_{21}}{a_{11}a_{22} - a_{21}a_{12}}. \tag{11.4}$$

Now let us rewrite the linear system of equations in matrix notation. Let

$$A = \begin{pmatrix} a_{11} & a_{12} \\ a_{21} & a_{22} \end{pmatrix} \tag{11.5}$$

be a 2×2 matrix of coefficients. Also let

$$X = \begin{pmatrix} x_1 \\ x_2 \end{pmatrix} \tag{11.6}$$

and

$$B = \begin{pmatrix} b_1 \\ b_2 \end{pmatrix} \tag{11.7}$$

be two 2×1 matrices (vectors). Then the linear system of equations can be written in compact form

$$AX = B. \tag{11.8}$$

In equation (11.8) you take the first row in A times X and equate to b_1 and you obtain the first of the linear equations (equation (11.1)).

Now fuzzify equation (11.8). $\overline{A}$ is a 2×2 matrix whose elements $\overline{a}_{ij}$ are continuous fuzzy numbers. $\overline{A}$ is called a Type II fuzzy matrix (Type I fuzzy matrices were studied in Chapter 7). Next let $\overline{X} = (\overline{X}_i)$ be a 2×1 matrix whose elements are $\overline{X}_1$ and $\overline{X}_2$ the unknown continuous fuzzy numbers. Also, $\overline{B} = (\overline{B}_i)$, another 2×1 matrix, with members $\overline{B}_1$ and $\overline{B}_2$ given continuous fuzzy numbers. The fuzzified equation (11.8) is

$$\overline{A} \cdot \overline{X} = \overline{B}, \tag{11.9}$$

for given $\overline{A}$ and $\overline{B}$ and we are to solve it for $\overline{X}$. Let $\overline{a}_{ij}[\alpha] = [a_{ij1}(\alpha), a_{ij2}(\alpha)]$ for $1 \leq i, j \leq 2$ and let $\overline{B}_j[\alpha] = [b_{j1}(\alpha), b_{j2}(\alpha)]$ for $j = 1, 2$ and set $\overline{X}_i[\alpha] = [x_{i1}(\alpha), x_{i2}(\alpha)]$, $i = 1, 2$.

We start with the classical solution, written $\overline{X}_c$, if it exists. In the classical solution we: (1) substitute α-cuts of $\overline{a}_{ij}, \overline{X}_i, \overline{B}_j$ for $\overline{a}_{ij}, \overline{X}_i, \overline{B}_j$, respectively; (2) simplify using interval arithmetic; (3) solve for $x_{ij}(\alpha)$, $1 \leq i, j \leq 2$; and (4) if the intervals $[x_{i1}(\alpha), x_{i2}(\alpha)]$, $i = 1, 2$, define continuous fuzzy numbers $\overline{X}_i$, $i = 1, 2$, then $\overline{X}_c$ exists and its components are $\overline{X}_1$ and $\overline{X}_2$. After substituting the α-cuts we obtain two interval equations

$$\begin{aligned} &[a_{111}(\alpha), a_{112}(\alpha)][x_{11}(\alpha), x_{12}(\alpha)] + \\ &\quad + [a_{121}(\alpha), a_{122}(\alpha)][x_{21}(\alpha), x_{22}(\alpha)] = \\ &= [b_{11}(\alpha), b_{12}(\alpha)], \end{aligned} \tag{11.10}$$

and

$$\begin{aligned} &[a_{211}(\alpha), a_{212}(\alpha)][x_{11}(\alpha), x_{12}(\alpha)] + \\ &\quad + [a_{221}(\alpha), a_{222}(\alpha)][x_{21}(\alpha), x_{22}(\alpha)] = \\ &= [b_{21}(\alpha), b_{22}(\alpha)], \end{aligned} \tag{11.11}$$

to simplify.

Now we need to know if the $\overline{a}_{ij}$ are positive or negative and know if the $\overline{X}_i$ are to be positive or negative. Assuming that all the $\overline{a}_{ij}$ and all the $\overline{B}_j$ are triangular fuzzy numbers, what we might do is first solve the crisp problem obtained from the ($\alpha = 1$)-cut. Suppose the crisp system produces solution $x_1 = 5$ and $x_2 = -7$. Then we would start by trying $\overline{X}_1 > 0$ whose ($\alpha = 1$)-cut is equal to 5 and $\overline{X}_2 < 0$ whose ($\alpha = 1$)-cut is -7.

Let us assume for this discussion that all the $\overline{a}_{ij} > 0$ and all the $\overline{B}_j > 0$ so that we first try (the crisp solution was positive) for $\overline{X}_i > 0$ all i. Then from equations (11.10) and (11.11) we get a 4×4 system

$$a_{111}(\alpha)x_{11}(\alpha) + a_{121}(\alpha)x_{21}(\alpha) = b_{11}(\alpha), \tag{11.12}$$

$$a_{112}(\alpha)x_{12}(\alpha) + a_{122}(\alpha)x_{22}(\alpha) = b_{12}(\alpha), \tag{11.13}$$

$$a_{211}(\alpha)x_{11}(\alpha) + a_{221}(\alpha)x_{21}(\alpha) = b_{21}(\alpha), \tag{11.14}$$

$$a_{212}(\alpha)x_{12}(\alpha) + a_{222}(\alpha)x_{22}(\alpha) = b_{22}(\alpha), \tag{11.15}$$

to solve for the $x_{ij}(\alpha)$, $1 \leq i,j \leq 2$, $0 \leq \alpha \leq 1$. We again assume that this 4×4 system has a unique solution for the $x_{ij}(\alpha)$, $1 \leq i,j \leq 2$, for all α.

After solving for the $x_{ij}(\alpha)$ we check to see if the intervals $[x_{i1}(\alpha), x_{i2}(\alpha)]$ define continuous fuzzy numbers for $i = 1,2$. What is needed is: (1) $\partial x_{i1}/\partial\alpha > 0$, (2) $\partial x_{i2}/\partial\alpha < 0$, and (3) $x_{i1}(1) \leq x_{i1}(1)$ (equality for triangular shaped fuzzy numbers), for $i = 1,2$.

Example 11.1

Let $\overline{a}_{11} = (4/5/7)$, $\overline{a}_{12} = 0$ (real number zero), $\overline{a}_{21} = 0$ and $\overline{a}_{22} = (6/8/12)$. Also set $\overline{B}_1 = (1/2/3)$ and $\overline{B}_2 = (2/5/8)$. Then $\overline{a}_{11}[\alpha] = [4+\alpha, 7-2\alpha]$, $\overline{a}_{22}[\alpha] = [6+2\alpha, 12-4\alpha]$, $\overline{B}_1[\alpha] = [1+\alpha, 3-\alpha]$ and $\overline{B}_2[\alpha] = [2+3\alpha, 8-3\alpha]$. Set $\alpha = 1$ and the crisp solution is $x_1 = 2/5$ and $x_2 = 5/8$ so we assume we can get a solution with $\overline{X}_1 > 0$, $\overline{X}[1] = 2/5$ and $\overline{X}_2 > 0$, $\overline{X}[1] = 5/8$.

Substitute these α-cuts into equations (11.12)–(11.15), using $\overline{a}_{12} = \overline{a}_{21} = 0$, and solve for the $x_{ij}(\alpha)$. We obtain

$$x_{11}(\alpha) = (1+\alpha)/(4+\alpha), \tag{11.16}$$

$$x_{12}(\alpha) = (3-\alpha)/(7-2\alpha), \tag{11.17}$$

$$x_{21}(\alpha) = (2+3\alpha)/(6+2\alpha), \tag{11.18}$$

$$x_{22}(\alpha) = (8-3\alpha)/(12-4\alpha). \tag{11.19}$$

We see that $\partial x_{11}/\partial\alpha > 0$, $\partial x_{12}/\partial\alpha < 0$, $\partial x_{21}/\partial\alpha > 0$ $\partial x_{22}/\partial\alpha < 0$, $x_{11}(1) = x_{12}(1)$ and $x_{21}(1) = x_{22}(1)$. Hence $\overline{X}_c$ exists with components $\overline{X}_1$ and $\overline{X}_2$ and the α-cuts of $\overline{X}_1$ are

$$[\frac{1+\alpha}{4+\alpha}, \frac{3-\alpha}{7-2\alpha}], \tag{11.20}$$

and the α-cuts of $\overline{X}_2$ are

$$[\frac{2+3\alpha}{6+2\alpha}, \frac{8-3\alpha}{12-4\alpha}]. \tag{11.21}$$

Example 11.2

Define $\overline{a}_{11} = (1/2/3)$, $\overline{a}_{12} = 0$, $\overline{a}_{21} = 1$ (crisp number one), $\overline{a}_{22} = (2/5/8)$, $\overline{B}_1 = (4/5/7)$ and $\overline{B}_2 = (6/8/12)$. Then $\overline{a}_{11}[\alpha] = [1+\alpha, 3-\alpha]$, $\overline{a}_{22}[\alpha] = [2+3\alpha, 8-3\alpha]$, $\overline{B}_1[\alpha] = [4+\alpha, 7-2\alpha]$ and $\overline{B}_2[\alpha] = [6+2\alpha, 12-4\alpha]$. If $\alpha = 1$, then the crisp solution is $x_1 = 5/2$ and $x_2 = 11/2$. So we expect a solution of the form $\overline{X}_1 > 0$ with $\overline{X}_1[1] = 5/2$ and $\overline{X}_2 > 0$ with $\overline{X}_2[1] = 11/2$.

Substitute the α-cuts into equations (11.12)–(11.15), using $\overline{a}_{12} = 0$ and $\overline{a}_{21} = 1$, and solve for the $x_{ij}(\alpha)$. We obtain

$$x_{11}(\alpha) = (4+\alpha)/(1+\alpha), \tag{11.22}$$

and

$$x_{12}(\alpha) = (7-2\alpha)/(3-\alpha). \tag{11.23}$$

But $\partial x_{11}/\partial\alpha < 0$. Hence $\overline{X}_c$ does not exist in this example.

The extension principle solution, written $\overline{X}_e$ with components $\overline{X}_1$ and $\overline{X}_2$, fuzzifies the crisp solution, usually using the t-norm min, given in equations (11.3) and (11.4). $\overline{X}_e$ always exists but it may, or may not, satisfy the original system of fuzzy linear equations. That is, $\overline{A} \cdot \overline{X}_e = \overline{B}$ may or may not be true. We check to see if $\overline{A} \cdot \overline{X}_e$ equals $\overline{B}$ using α-cuts and interval arithmetic.

Let $h_1(a_{11}, \cdots, a_{22}, b_1, b_2)$ be the expression equal to x_1 in equation (11.3) and set $h_2(a_{11}, \cdots, a_{22}, b_1, b_2)$ to be the formula equal to x_2 in equation (11.4). To find $\overline{X}_1$ in $\overline{X}_e$ we substitute $\overline{a}_{11}, \cdots, \overline{B}_2$ for $a_{11}, \cdots b_2$ in $h_1(a_{11}, \cdots, b_2)$ and evaluate using the extension principle. We now need to assume that the denominators in equations (11.3) and (11.4) are never zero. So we assume that $a_{11}a_{22} - a_{21}a_{12} \neq 0$ for all $a_{ij} \in \overline{a}_{ij}[0]$ for $1 \leq i, j \leq 2$.

We find the α-cuts of $\overline{X}_1$ as

$$x_{11}(\alpha) = \min\{h_1(a_{11}, \cdots, b_2) | a_{ij} \in \overline{a}_{ij}[\alpha], b_j \in \overline{B}_j[\alpha]\}, \tag{11.24}$$

$$x_{12}(\alpha) = \max\{h_1(a_{11}, \cdots, b_2) | a_{ij} \in \overline{a}_{ij}[\alpha], b_j \in \overline{B}_j[\alpha]\}, \tag{11.25}$$

for $0 \leq \alpha \leq 1$. Similarly we find α-cuts of $\overline{X}_2$ using $h_2(a_{11}, \cdots, b_2)$. Set $\overline{X}_e = (\overline{X}_i)$ and check to see if $\overline{A} \cdot \overline{X}_e = \overline{B}$.

In general, equations (11.24) and (11.25) will be difficult to evaluate. One may use a genetic algorithm (Chapter 15) to estimate $x_{1i}(\alpha)$ for selected values of α. Notice that if you set $\alpha = 1$ in equations (11.24) and (11.25), then you get, assuming that all the $\overline{a}_{ij}$ and $\overline{B}_j$ are triangular shaped fuzzy numbers, $x_{11}(1) = x_{12}(1) = x_1$ in the crisp solution. Similarly $\alpha = 1$ gives x_2 in the crisp solution using $h_2(a_{11}, \cdots, b_2)$.

Example 11.3

This continues Example 11.1 using the same $\overline{a}_{ij}$ and $\overline{B}_j$. Now $h_1(a_{11},\cdots,b_2) = b_1/a_{11}$ and $h_2(a_{11},\cdots,b_2) = b_2/a_{22}$. We easily see that using the extension principle we get

$$[x_{11}(\alpha), x_{12}(\alpha)] = [\frac{1+\alpha}{7-2\alpha}, \frac{3-\alpha}{4+\alpha}], \tag{11.26}$$

$$[x_{21}(\alpha), x_{22}(\alpha)] = [\frac{2+3\alpha}{12-4\alpha}, \frac{8-3\alpha}{6+2\alpha}]. \tag{11.27}$$

Notice that $\overline{X}_1[1] = 2/5$ and $\overline{X}_2[1] = 5/8$ and $\overline{X}_c \leq \overline{X}_e$.

Example 11.4

This continues Example 11.2 with the same values for the $\overline{a}_{ij}$ but we change $\overline{B}_1 = (1.5/2.5/3.5)$ and $\overline{B}_2 = (10/12/14)$ because it will be easier to find $\overline{X}_e$ using these new values for $\overline{B}_j$. We find $h_1(a_{11},\cdots,b_2) = b_1/a_{11}$ as before but $h_2(a_{11},\cdots,b_2) = (a_{11}b_2 - b_1)/(a_{11}a_{22})$. We find $\overline{X}_1$ as in Example 11.3. To find α-cuts of $\overline{X}_2$ we need to evaluate equations (11.24) and (11.25) using h_2. We find that h_2 is: (1) an increasing function of a_{11} and b_2, but (2) it is a decreasing function of a_{22} and b_1. Hence

$$x_{21}(\alpha) = h_2(a_{111}(\alpha), a_{222}(\alpha), b_{12}(\alpha), b_{21}(\alpha)), \tag{11.28}$$

$$x_{22}(\alpha) = h_2(a_{112}(\alpha), a_{221}(\alpha), b_{11}(\alpha), b_{22}(\alpha)), \tag{11.29}$$

for $0 \leq \alpha \leq 1$. The interval $[x_{21}(\alpha), x_{22}(\alpha)]$ defines α-cuts of $\overline{X}_2$. The extension principle solution $\overline{X}_e = (\overline{X}_i)$.

We note that $\overline{X}_c$ does not exist for these new values of the $\overline{B}_j$.

The last solution type, written $\overline{X}_I$, substitutes α-cuts of the $\overline{a}_{ij},\cdots,\overline{B}_2$ for $a_{11},\cdots,b_2$, respectively, into h_1 and h_2, simplifies using interval arithmetic, giving α-cuts of the components of $\overline{X}_I$. If $\overline{A}\cdot\overline{X}_I = \overline{B}$, using α-cuts and interval arithmetic, then we say that $\overline{X}_I$ satisfies the system of fuzzy linear equations. We have changed notation in that we now write $\overline{X}_I$ for the α-cut and interval arithmetic solution whereas in Chapter 5 we used $\overline{X}_i$ for this solution. In this chapter we are using $\overline{X}_i$ for the components of the vector $\overline{X}$.

Example 11.5

This continues Examples 11.1 and 11.3 with the same $\overline{a}_{ij}$ and $\overline{B}_j$. We compute

$$\overline{X}_1[\alpha] = \overline{B}_1[\alpha]/\overline{a}_{11}[\alpha], \tag{11.30}$$

$$\overline{X}_2[\alpha] = \overline{B}_2[\alpha]/a_{22}[\alpha]. \tag{11.31}$$

This defines the α-cuts of $\overline{X}_1$ and $\overline{X}_2$ the two components of $\overline{X}_I$. We see that $\overline{X}_e = \overline{X}_I$.

Example 11.6

This continues Example 11.4. Since $h_1(a_{11}, \cdots, b_2) = b_1/a_{11}$ we get the same value for $\overline{X}_1$, using α-cuts and interval arithmetic, as the extension principle (equation (11.26)). For $\overline{X}_2$ we obtain

$$\overline{X}_2[\alpha] = (\overline{a}_{11}[\alpha]\overline{B}_2[\alpha] - \overline{B}_1[\alpha])/(\overline{a}_{11}[\alpha]\overline{a}_{22}[\alpha]). \tag{11.32}$$

All multiplications are easy because all the fuzzy numbers are positive. The numerator in equation (11.32) is $[N_1(\alpha), N_2(\alpha)]$ where

$$N_1(\alpha) = a_{111}(\alpha)b_{21}(\alpha) - b_{12}(\alpha), \tag{11.33}$$

and

$$N_2(\alpha) = a_{112}(\alpha)b_{22}(\alpha) = b_{11}(\alpha). \tag{11.34}$$

The denominator is $[D_1(\alpha), D_2(\alpha)]$ which is

$$D_1(\alpha) = a_{111}(\alpha)a_{221}(\alpha), \tag{11.35}$$

and

$$D_2(\alpha) = a_{112}(\alpha)a_{222}(\alpha). \tag{11.36}$$

Then

$$\overline{X}_2[\alpha] = [\frac{N_1(\alpha)}{D_2(\alpha)}, \frac{N_2(\alpha)}{D_1(\alpha)}]. \tag{11.37}$$

For example

$$x_{21}(\alpha) = \frac{a_{111}(\alpha)b_{21}(\alpha) - b_{12}(\alpha)}{a_{112}(\alpha)a_{222}(\alpha)}. \tag{11.38}$$

For $\overline{X}_e$ and $\overline{X}_I$ we see that they have the same first component $\overline{X}_1$ but not the same second component $\overline{X}_2$. For $\overline{X}_e$ we have $\overline{X}_2[0] = [0.81, 6.75]$ and $\overline{X}_2[1] = 2.15$. For $\overline{X}_I$ we get $\overline{X}_2[0] = [0.27, 20.25]$ and $\overline{X}_2[1] = 2.15$.

In general, when $\overline{X}_c$ exists we obtain $\overline{X}_c \leq \overline{X}_e \leq \overline{X}_I$. If the components of $\overline{X}_c(\overline{X}_e, \overline{X}_I)$ are $\overline{X}_{cj}(\overline{X}_{ej}, \overline{X}_{Ij})$, for $j = 1, 2$, then the above inequality means that $\overline{X}_{cj} \leq \overline{X}_{ej} \leq \overline{X}_{Ij}$ for $j = 1, 2$. The fuzziness grows as you go from $\overline{X}_c$ to $\overline{X}_I$. Fuzziness is measured by the length of the support (or base) and the length of the core for trapezoidal (shaped) fuzzy numbers.

11.1 Exercises

1. If $\overline{a}_{11} > 0$, $\overline{a}_{12} > 0$, $\overline{a}_{21} < 0$, $\overline{a}_{22} > 0$, $\overline{B}_1 > 0$, $\overline{B}_2 > 0$, $\overline{X}_1 < 0$ and $\overline{X}_2 > 0$, then write down the new equations corresponding to equations (11.12)–(11.15) that are needed to find the $x_{ij}(\alpha)$, $1 \leq i, j \leq 2$.

2. In Example 11.1 is it possible to have other fuzzy solutions for $\overline{X}_1 < 0$ and/or $\overline{X}_2 < 0$? Explain.

3. Answer the question in Problem 2 also for Example 11.2.

4. Let $\overline{a}_{11} = \overline{a}_{22} = 0$ and $\overline{a}_{12} = (4/5/7)$, $\overline{a}_{21} = (6/8/12)$, $\overline{B}_1 = (-3/-2/-1)$ and $\overline{B}_2 = (2/5/8)$. Find α-cuts for:

   a. $\overline{X}_c$, or show it does not exist;

   b. $\overline{X}_e$ and show $\overline{X}_c \leq \overline{X}_e$, if $\overline{X}_c$ exists ;

   c. $\overline{X}_I$ and show that $\overline{X}_e \leq \overline{X}_I$.

5. Let $\overline{a}_{11} = 0$, $\overline{a}_{12} = 1$, $\overline{a}_{21} = (1/2/3)$, $\overline{a}_{22} = (2/4/8)$, $\overline{B}_1 = 3$ and $\overline{B}_2 = (7/14/27)$. Find α-cuts for:

   a. $\overline{X}_c$, or show that it does not exist;

   b. $\overline{X}_e$ and show that $\overline{X}_c \leq \overline{X}_e$, when $\overline{X}_c$ exists;

   c. $\overline{X}_I$ and show that $\overline{X}_e \leq \overline{X}_I$.

6. Show that, in general, if $\overline{X}_c$ exists, then $\overline{X}_c \leq \overline{X}_e$.

7. Show that, in general, $\overline{X}_e \leq \overline{X}_I$.

8. In Examples 11.1 and 11.3 show that $\overline{X}_c \leq \overline{X}_e$.

9. Using the values of $\overline{B}_1$ and $\overline{B}_2$ given in Example 11.4 show that $\overline{X}_c$ does not exist.

10. Assuming that $a_{11}a_{22} - a_{12}a_{21} \neq 0$, then A^{-1} exists ($A^{-1}A = I$ the 2×2 identity). Use the crisp solution formula

$$X = A^{-1}B$$

    to compute both $\overline{X}_e$ and $\overline{X}_I$. Do you get the same results as using h_1 and h_2 as was done in the text? Explain.

11. Suppose we have values of the $\overline{a}_{ij}$ so that $a_{11}a_{22} - a_{12}a_{21} = 0$ for certain values of a_{ij} in $\overline{a}_{ij}[\alpha]$, for some α in $[0, 1)$. Discuss the problems we now face in defining and computing both $\overline{X}_e$ and $\overline{X}_I$.

12. Determine if $\overline{X}_e$ in Example 11.3 satisfies the original system of fuzzy linear equations.

13. Same as Problem 12 except for $\overline{X}_e$ in Example 11.4.

14. Same as Problem 12 except for $\overline{X}_I$ in Example 11.6.

15. Let $\overline{a}_{12} = \overline{a}_{21} = 1$, $\overline{a}_{11} = (4/5/7)$, $\overline{a}_{22} = (6/8/12)$, $\overline{B}_1 = (1/2/3)$ and $\overline{B}_2 = (2/5/8)$. Find α-cuts for :

    a. $\overline{X}_c$, if it exists;

    b. $\overline{X}_e$;

    c. $\overline{X}_I$.

16. In defining $\overline{X}_e$ we used the t-norn min in the extension principle. Use another t-norm T_b, or T_p or T^* to find $\overline{X}_e$ in Example 11.3. Also try another t-norm in Example 11.4.

17. Fuzzy Eigenvalues. A fuzzy eigenvalue is a continuous fuzzy number $\overline{\lambda}$ satisfying the equation

$$\overline{A} \cdot \overline{X} = \overline{\lambda} \cdot \overline{X},$$

    for $\overline{X} \neq 0$. Discuss the definition of, and computation for:

    a. $\overline{\lambda}_c$;

    b. $\overline{\lambda}_e$;

    c. $\overline{\lambda}_i$ (the α-cut and interval arithmetic solution).

18. Apply the results of Problem 17 to Example 11.1. That is, use the data in this example to calculate the fuzzy eigenvalues.

19. Consider the case where $\overline{X}_c$ does not exist and then substitute classical solution $\overline{X}_s$ defined in Problem 10 in Section 5.2.4. $\overline{X}_s$ is that triangular shaped fuzzy number that solves the following minimization problem:

$$\min D(\overline{A} \cdot \overline{X}, \overline{B}).$$

    Start with the D given in equation (3.80) in Section 3.7 and extend it to a D for finding the distance between $\overline{A}\,\overline{X}$ and $\overline{B}$. Discuss a method for solving for $\overline{X}_s$.

20. Apply the results of Problem 19 to Example 11.2 and compute $\overline{X}_s$. Compare your result to $\overline{X}_e$ found in Example 11.4.

Chapter 12

Possibility Theory

12.1 Introduction

General possibility theory would be part of measure theory, fuzzy measures and evidence theory, and as such is beyond the scope of this book and to be included in a more advanced course. Instead we will introduce discrete possibilities in the next section and compare their properties to discrete probabilities. As an application of discrete possibilities we discuss finite Markov chain, based on possibility theory, in the third section.

12.2 Discrete Possibilities

In this section we will introduce discrete possibility theory and discuss some of its properties. We will also compare to discrete probability theory to see the similarities and differences between the two theories.

Let $X = \{x_1, x_2, \cdots, x_n\}$ be a finite set. *Poss* will stand for possibility and *Prob* will denote probability. A discrete possibility distribution on X is

$$Poss = \{\frac{\mu_1}{x_1}, \cdots, \frac{\mu_n}{x_n}\}, \tag{12.1}$$

where the μ_i are in the interval $[0, 1]$ and there is at least one μ_i equal to one (basic constraint of possibilities).

A discrete possibility distribution on X is a normalized fuzzy subset of X. A discrete probability distribution on X is

$$Prob = \{\frac{p_1}{x_1}, \cdots, \frac{p_n}{x_n}\}, \tag{12.2}$$

where the p_i are in the interval $[0, 1]$ but the basic constraint on probabilities is

$$p_1 + p_2 + \cdots + p_n = 1. \tag{12.3}$$

Let **U** be a set and f a function mapping **U** into X. f is called a fuzzy variable if its values are restricted by a possibility distribution. What this means is that if E is a crisp subset of X, and *Poss* in equation (12.1) is the restricting possibility distribution, then

$$Poss[f(u) \in E] = \max\{\mu_i | f(u) = x_i \in E\}, \tag{12.4}$$

where u is in **U**. The possibility that f takes on a value in E is the maximum of the possibilities of all the x in E for which there is an $u \in \mathbf{U}$ so that $f(u) = x \in E$. A function g is called a random variable, g also maps **U** into X, if the values of g are restricted by a probability distribution.

Assuming that *Prob* in equation (12.2) is the probability distribution restricting the values of g, then

$$Prob[g(u) \in E] = \sum\{p_i | g(u) = x_i \in E\}. \tag{12.5}$$

Here we sum the relevant probabilities. This is the usual analogue between fuzzy and crisp: use max in fuzzy and addition in non-fuzzy.

Now let A and B be two crisp subsets of X. Then

$$Poss[A \cup B] = \max[Poss(A), Poss(B)], \tag{12.6}$$

but

$$Prob[A \cup B] = Prob(A) + Prob(B) - Prob(A \cap B), \tag{12.7}$$

which is

$$Prob[A \cup B] = Prob(A) + Prob(B), \tag{12.8}$$

when $A \cap B = \phi$.

We can also easily see that

$$Poss(A) + Poss(A^c) \geq 1, \tag{12.9}$$

$$\max(Poss(A), Poss(A^c)) = 1, \tag{12.10}$$

but

$$Prob(A) + Prob(A^c) = 1. \tag{12.11}$$

Total ignorance in possibility theory means that $\mu_i = 1$ for all i in equation (12.1), but in probability theory total ignorance can mean $p_i = \frac{1}{n}$ for all i, called the uniform probability distribution.

Next we consider joint distributions. Let $Y = \{y_1, \cdots, y_m\}$ be another finite set with

$$Poss = \{\frac{\theta_1}{y_1}, \cdots, \frac{\theta_m}{y_m}\}], \tag{12.12}$$

$$Prob = \{\frac{q_1}{y_1}, \cdots, \frac{q_m}{y_m}\}. \tag{12.13}$$

We also have function h (k) mapping set **V** into Y whose values are restricted by Poss (Prob) given in equation (12.12) [(12.13)]. Assuming the fuzzy variables f and h are non-interactive (a possibility theory term), then their joint possibility distribution is given by

$$Poss[f(u) = x_i \text{ and } h(v) = y_j] = \min\{\mu_i, \theta_j\}, \tag{12.14}$$

$1 \leq i \leq n$ and $1 \leq j \leq m$.

But if random variables g and k are independent, their joint probability distribution is

$$Prob[g(u) = x_i \text{ and } k(v) = y_j] = p_i q_j, \tag{12.15}$$

for $1 \leq i \leq n$ and $1 \leq j \leq m$. Another difference between the two theories is in the use of min for multiplication in the fuzzy case.

12.2.1 Exercises

All problems involve possibility theory. No problems use probabilities.

1. Show $Poss(A \cap B) \leq \min\{Poss(A), Poss(B)\}$.

2. Show $Poss(A) + Poss(A^c) \geq 1$.

3. Show $\max(Poss(A), Poss(A^c)) = 1$.

4. Let

$$Poss = \{\frac{0.6}{0}, \frac{1}{1}, \frac{0.5}{2}, \frac{0.3}{3}, \frac{0.2}{4}\},$$

be a discrete possibility distribution on $X = \{0, 1, 2, 3, 4\}$. For every subset E of X, and there are 32 such subsets, find $Poss[E]$.

5. Let

$$Poss = \{\frac{0.2}{a}, \frac{1}{e}, \frac{0.7}{i}, \frac{1}{o}, \frac{0.6}{u}, \frac{0.5}{y}\},$$

be a possibility distribution on $X = \{a, e, i, o, u, y\}$. Let $\mathbf{U} = \{0, 1, 2, 3, 4\}$ and $f(0) = i, f(1) = u, f(2) = a, f(3) = y, f(4) = o$. Find the value of $Poss[f(u) \in E]$ if

a. $E = \{e\}$.

b. $E = \{a, u, y\}$.

c. $E = \{i, e, y\}$.

Find the value of $Poss[f(u) \in (A \cup B)]$ if

d. $A = \{a, e\}$ and $B = \{i, e, u\}$.

e. $A = \{i, o, u\}$ and $B = \{a, e\}$.

The following problems have to do with $\overline{E}$ a fuzzy subset of $X \times Y$ for $X = \{x_1, \cdots, x_n\}$ and $Y = \{y_1, \cdots, y_m\}$. $\overline{E}$ will be a joint possibility distribution on $X \times Y$ if it is normalized (membership value one for some $(x_i, y_j) \in X \times Y$). Assume we have

$$\overline{E}(x_i, y_j) = Poss[f(u) = x_i \text{ and } h(v) = y_j],$$

for fuzzy variables $f : \mathbf{U} \to X$ and $h : \mathbf{V} \to Y$. The marginal possibility distributions are calculated as follows

$$Poss_x[x_i] = Poss[f(u) = x_i] = \max_y \overline{E}(x_i, y),$$

$$Poss_y[y_j] = Poss[h(v) = y_j] = \max_x \overline{E}(x, y_j).$$

The fuzzy variables f and h, or the possibility distributions $Poss_x$ and $Poss_y$, are said to be non-interactive, if

$$\overline{E}(x_i, y_j) = \min\{Poss_x[x_i], Poss_y[y_j]\},$$

for $1 \le i \le n$ and $1 \le j \le m$.

6. Let $\overline{E}$ be given by

$\overline{E}$	y_1	y_2	y_3
x_1	0	0.7	0.6
x_2	1	0.4	0.2
x_3	0.9	1	0.3
x_4	0.5	1	0.8

   a. If $A = \{(x_2, y_2), (x_2, y_3), (x_4, y_1)\}$, then find $Poss[A]$.
   b. Find $Poss_x$ and show that it is a possibility distribution (it is normalized).
   c. Find $Poss_y$ and show that it is normalized.
   d. Are $Poss_x$ and $Poss_y$ non-interactive? Explain your answer.

7. Let $\overline{E}$ be

$\overline{E}$	y_1	y_2	y_3	y_4	y_5
x_1	0.2	0.3	0.3	0.3	0.3
x_2	0.2	0.3	0.5	0.7	0.7
x_3	0.2	0.3	0.5	0.7	1
x_4	0.2	0.3	0.5	0.6	0.6
x_5	0.2	0.2	0.2	0.2	0.2

   a. If $A = \{(x_3, y_4), (x_4, y_2), (x_5, y_5), (x_2, y_3), (x_1, y_1)\}$, then find $Poss[A]$.
   b. Find $Poss_x$. When is $Poss_x[x_i] = 1$?
   c. Find $Poss_y$.
   d. Are $Poss_x$ and $Poss_y$ non-interactive? Explain your answer.

12.3 Fuzzy Markov Chains

We first review some of the basic results of finite Markov chains based on probability theory and then we present fuzzy finite Markov chains based on possibility theory. We look at convergence of powers of the transition matrix and absorbing Markov chains for fuzzy Markov chains. We present some of the basic results of finite Markov chains based on probability theory so that we may see the similarities and differences between the two theories.

A finite Markov chain has a finite number of possible states (outcomes) $S_1, \cdots, S_r$ at each step $n = 1, 2, 3, \cdots$ in the process. Let

$$p_{ij} = Prob\{S_j \ at \ step \ n+1 | S_i \ at \ step \ n\}, \tag{12.16}$$

$1 \leq i, j \leq r$ and $n = 1, 2, 3, \cdots$. The p_{ij} are the transition probabilities which do not depend on n. The transition matrix $P = (p_{ij})$ is a $r \times r$ matrix of the transition probabilities. An important property of P is that the row sums are equal to one (each row is a discrete probability distribution) and each $p_{ij} \geq 0$. Let $p_{ij}^{(n)}$ be the probability of starting off in state S_i and ending in state S_j after n steps. Define P^n to be the product of P n-times and it is well known that $P^n = (p_{ij}^{(n)})$ for all n. If $p^{(0)} = (p_1^{(0)}, \cdots, p_r^{(0)})$, where $p_i^{(0)} =$ the probability of initially being in state S_i, and let $p^{(n)} = (p_1^{(n)}, \cdots, p_r^{(n)})$, where $p_i^{(n)} =$ the probability of being in state S_i after n steps, then we know that $p^{(n)} = p^{(0)} P$.

We say that the Markov chain is regular if $P^k > 0$ for some positive integer k, which is $p_{ij}^{(k)} > 0$ for all i and j. This means it is possible to go from any state S_i to any state S_j in k steps. A property of regular Markov chains is that powers of P converge, or $\lim_{n \to \infty} P^n = \Pi$, where the rows of Π are all identical. Each row in Π is equal to some $w = (w_1, \cdots, w_r)$ and $p^{(n)} \to p^{(0)}\Pi = w$. After a long time, thinking that each step takes a certain amount of time, the probability of being in state S_i is w_i, $1 \leq i \leq r$, independent of the initial conditions given in $p^{(0)}$. In a regular Markov chain the process goes on forever jumping from state to state.

We will call a state S_i absorbing if $p_{ii} = 1$ and $p_{ij} = 0$ for $j \neq i$. Once you enter S_i you can never leave. Suppose there are k absorbing states, $1 \leq k < r$, and then we may rename the states (if needed) so that the transition matrix P can be written as

$$P = \begin{pmatrix} I & O \\ R & Q \end{pmatrix} \tag{12.17}$$

where I is a $k \times k$ identity matrix, O is the $k \times (r-k)$ zero matrix, R is $(r-k) \times k$ and Q is a $(r-k) \times (r-k)$ matrix. The Markov chain is called an absorbing Markov chain if it has at least one absorbing state and from every non-absorbing state it is possible to reach some absorbing state in a finite number of steps. Assuming the Markov chain is absorbing we then know that

$$P^n = \begin{pmatrix} I & O \\ SR & Q^n \end{pmatrix} \tag{12.18}$$

where $S = I + Q + \cdots + Q^{n-1}$. Then $\lim_{n\to\infty} P^n = \Pi$ where

$$\Pi = \begin{pmatrix} I & O \\ R^* & O \end{pmatrix} \tag{12.19}$$

for $R^* = (I - Q)^{-1}R$. The notation $(I - Q)^{-1}$ is for the matrix inverse of $(I - Q)$, or their product is the identity matrix I. Notice the zero columns in Π which implies that the probability that the process will eventually enter an absorbing state in one. The process eventually ends up in an absorbing state.

Now we can introduce fuzzy Markov chains based on possibility theory. We have the same finite number of states $S_1, \cdots, S_r$ but now p_{ij} = the possibility that the process is in state S_j at step $n + 1$ given it was in state S_i at step n, $1 \leq i, j \leq r$. The p_{ij} are now the transition possibilities independent of n. Let $P = (p_{ij})$ a $r \times r$ matrix of possibilities, where now each row maximun is one, needed for possibility distributions. Let $p_{ij}^{(n)}$ = the possibility of being in state S_j at step n, given you started in state S_i. Let P^n be the n fold produce of P using the max-min composition of fuzzy matrices (relations). That is we use min in place of multiplication and max for addition. We claim that $P^n = (p_{ij}^{(n)})$. Our only assumption is that all the possibility distributions are non-interactive. Non-interactive for possibility distributions is analogous to assuming random variables are independent in probability theory. If $p^{(0)}$ is the initial possibility distribution and if $p^{(n)} = (p_1^{(n)}, \cdots, p_r^{(n)})$, where $p_i^{(n)}$ = the possibility of being in state S_i after n steps, we also see that $p^{(n)} = p^{(0)}P^n$ for the max-min composition of $p^{(0)}$ and P^n.

It is known that the sequence P^n for $n = 1, 2, 3, \cdots$ either converges or oscillates (see Section 7.3). If P^n converges, then we will call the fuzzy Markov chain regular. If P^n oscillates, and does not converge, it will be called an oscillating fuzzy Markov chain. We will now study absorbing fuzzy Markov chains.

Just like for non-fuzzy absorbing Markov chains we rename the states, if needed, to get P as in equation (12.17). A state S_i is absorbing if $p_{ii} = 1$ and $p_{ij} = 0$ for $j \neq i$. Once in an absorbing state S_i, the possibility of leaving is zero. We assume there are k absorbing states for $1 \leq k < r$. We can show that

$$P^n = \begin{pmatrix} I & O \\ R_n & Q^n \end{pmatrix} \tag{12.20}$$

where $R_n = \max\{R, QR, \cdots, Q^{n-1}R\}$.

The sequence P^n, $n = 1, 2, 3 \cdots$ converges, or oscillates, depending on the sequence Q^n, $n = 1, 2, 3 \cdots$. If Q^n converges to C, then P^n converges because R_n always converges to R^* (see below). When $Q^n \to C$, than $P^n \to \Pi$ where

$$\Pi = \begin{pmatrix} I & O \\ R^* & C \end{pmatrix} \tag{12.21}$$

and C need not be the zero matrix. Now $R^* = \max\{R, QR, \cdots, Q^c R\}$ for some positive integer c. We always have $R_n \to R^*$ if Q^n converges or oscillates. If $Q^n \to O$, the zero matrix, then we will call this fuzzy Markov chain an absorbing fuzzy Markov chain.

Example 12.3.1

Let

$$P = \begin{pmatrix} 1 & 0 & 0 & 0 \\ 0 & 1 & 0 & 0 \\ 0 & 0.3 & 0.7 & 1 \\ 1 & 0.6 & 0.2 & 0 \end{pmatrix} \tag{12.22}$$

Now

$$Q = \begin{pmatrix} 0.7 & 1 \\ 0.2 & 0 \end{pmatrix} \tag{12.23}$$

so Q converges to C with $Q^2 = Q^3 = \cdots = C$ for

$$C = \begin{pmatrix} 0.7 & 0.7 \\ 0.2 & 0.2 \end{pmatrix} \tag{12.24}$$

We therefore see that $R^* = \max\{R, QR, Q^2 R\}$ and we calculate

$$R^* = \begin{pmatrix} 1 & 0.6 \\ 1 & 0.6 \end{pmatrix} \tag{12.25}$$

Hence $P^n \to \Pi$ and

$$\Pi = \begin{pmatrix} 1 & 0 & 0 & 0 \\ 0 & 1 & 0 & 0 \\ 1 & 0.6 & 0.7 & 0.7 \\ 1 & 0.6 & 0.2 & 0.2 \end{pmatrix} \tag{12.26}$$

We have $P^n = \Pi$ for $n \geq 3$. It is not an absorbing fuzzy Markov chain.

Example 12.3.2

If

$$R = \begin{pmatrix} 0.5 & 1 \\ 1 & 0.6 \end{pmatrix} \tag{12.27}$$

and

$$Q = \begin{pmatrix} 0 & 0.2 \\ 0 & 0 \end{pmatrix} \tag{12.28}$$

then $Q^2 = Q^3 = \cdots = C$ is the zero matrix, and P is the transition matrix for an absorbing fuzzy Markov chain.

12.3.1 Exercises

All problems deal only with fuzzy Markov chains.

1. Let

$$P = \begin{pmatrix} 0 & 0.2 & 1 \\ 0.4 & 0 & 1 \\ 0 & 1 & 0.3 \end{pmatrix}$$

If P is the transition matrix for a fuzzy Markov chain, is it a regular or oscillating fuzzy Markov chain?

2. Let

$$P = \begin{pmatrix} 1 & 0.2 & 0 \\ 0 & 1 & 0.2 \\ 0 & 0.2 & 1 \end{pmatrix}$$

Will this produce a regular or oscillating fuzzy Markov chain?

3. Let

$$P = \begin{pmatrix} 0.2 & 0.6 & 1 \\ 1 & 0 & 0.5 \\ 0.2 & 1 & 0.4 \end{pmatrix}$$

Will this be a regular or oscillating fuzzy Markov chain?

4. Let

$$P = \begin{pmatrix} 1 & 0 & 0 & 0 \\ 0.4 & 1 & 0.6 & 0 \\ 0 & 1 & 0.5 & 0 \\ 0 & 0 & 1 & 0 \end{pmatrix}$$

This is the transition matrix for a fuzzy Markov chain with one absorbing state. Does this transition matrix produce an absorbing fuzzy Markov chain?

5. Let

$$P = \begin{pmatrix} 1 & 0 & 0 & 0 & 0 \\ 0 & 1 & 0 & 0 & 0 \\ 0.7 & 0 & 1 & 0.1 & 0 \\ 0.5 & 1 & 0 & 0 & 0.2 \\ 0.2 & 0 & 0.5 & 1 & 0 \end{pmatrix}$$

Will this give an absorbing fuzzy Markov chain?

6. Give an argument that $R_2 = \max\{R, QR\}$ in equation (12.20).

7. Give an argument that R_n will eventually equal $\max\{R, QR, \cdots, Q^c R\}$ for some positive integer c.

8. Show that P^2, found using the max-min composition of P and P, gives an $r \times r$ matrix whose elements $p_{ij}^{(2)}$ compute the possibility of going from state S_i to state S_j in two steps.

9. Let P be the transition matrix for a fuzzy Markov chain. Show that each row in P^2 is a possibility distribution (maximum value of one).

10. Let

$$P = \begin{pmatrix} a_{11} & a_{12} \\ a_{21} & a_{22} \end{pmatrix}$$

for: (1) $a_{11} = a_{22} = 1$, (2) $a_{11} = a_{21} = 1$, (3) $a_{12} = a_{21} = 1$, and (4) $a_{12} = a_{22} = 1$. In all cases the other a_{ij} are arbitrary except their values must be in $[0, 1]$. The four cases give all possible 2×2 fuzzy transition matrices. Determine when P^n converges and when it oscillates.

11. Let P be the 2×2 transition matrix in Problem 10 but now we have the cases: (1) $a_{11} = a_{22} = 1$, $a_{12} = 0$ and a_{21} being an arbitrary number in $[0, 1]$, and (2) $a_{11} = a_{21} = 1$, $a_{12} = 0$ and a_{22} being an arbitrary number in $[0, 1]$. These two cases give all possible 2×2 fuzzy Markov chains with one absorbing state. Find P^n and determine when P^n converges and when we have an absorbing fuzzy Markov chain.

12. Write down all possible 3×3 transition matrices that have two absorbing states. Find P^n and when it converges. If it converges, then find the limit and when it will give an absorbing fuzzy Markov chain.

13. Let P be a transition matrix for a fuzzy Markov chain. Find a necessary and sufficient condition on P so that P^n converges (will not oscillate).

14. Let P be a transition matrix for a regular fuzzy Markov chain. Is P^n also a transition matrix for a regular fuzzy Markov chain?

15. Let P be a transition matrix for an absorbing fuzzy Markov chain. Is P^n also a transition matrix for an absorbing fuzzy Markov chain?

16. Let P be the transition matrix for a fuzzy Markov chain. P may also be considered a fuzzy relation. Decide if the following statements are true or false:

    a. If P is reflexive, then P is the transition matrix for a regular fuzzy Markov chain.

    b. If P is transitive, the P is the transition matrix for a regular fuzzy Markov chain.

Chapter 13

Neural Nets

13.1 Introduction

We first introduce layered, feedforward, neural nets in the next section and go through all the details on how they compute their output given inputs. Our applications of these neural nets is to: (1) approximate solutions (the α-cut and interval arithmetic solution of Section 5.2.3 in Chapter 5) to fuzzy equations, and (2) approximate the values (the α-cut and interval arithmetic value of Section 8.3 in Chapter 8) of fuzzy functions. In both cases the neural net requires sign constraints on its weights (some weights must be positive and the rest must be negative). Then in the third section of this chapter we fuzzify to get a fuzzy neural net. Our applications of fuzzy neural nets is to construct hybrid fuzzy neural nets for fuzzy functions. There is now no approximation, the output from the hybrid fuzzy neural nets will exactly equal the values of the fuzzy function.

13.2 Layered, Feedforward, Neural Nets

In this section we will first introduce layered, feedforward, neural nets (abbreviated simply as "neural net") and then show how they compute outputs from inputs. Then we show how certain neural nets can be used to approximate fuzzy functions and also approximate solutions to fuzzy equations.

A simple "2-3-1" neural net is shown in Figure 13.1. The notation "2-3-1" means two input neurons, 3 neurons in the second layer and one neuron in the output layer. We will usually use three layers in this chapter.

We first need to explain how the neural net computes its output y from its inputs x_1 and x_2. Real numbers x_1 and x_2 are sent to the input neurons. The input neurons simply distribute the x_i over the directed arcs shown in Figure 13.1 to the neurons in the second layer. All neurons have a transfer function and a shift term. The transfer function is represented by "f" inside

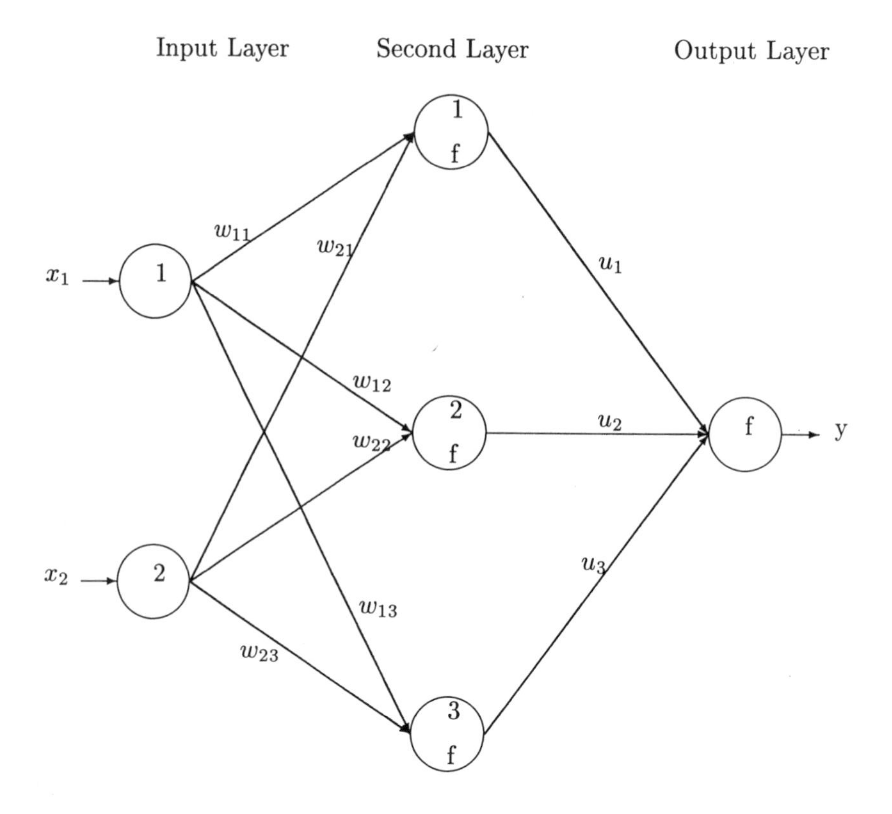

Figure 13.1: A 2-3-1 Neural Net

the neuron. If there is no "f" inside a neuron, then the transfer function is the identity function $i(x) = x$ for all x. Neurons using the identity function, only the input neurons in Figure 13.1, have no shift term.

On each arc, connecting an input neuron to a neuron in the second layer, is a real number weight w_{ij}. The neural net multiplies the signal x_i and the weight w_{ij}, then adds these results over all incoming arcs, and this sum is the input to that neuron in the second layer. For example, the input to neuron #2 in the second layer is

$$x_1 w_{12} + x_2 w_{22}. \tag{13.1}$$

Each neuron in the second layer adds a shift term θ_i and inputs the result into the transfer function. The result, the value of the transfer function, is the output from that neuron. For example, the output from neuron #2 in the second layer is

$$z_2 = f(x_1 w_{12} + x_2 w_{22} + \theta_2). \tag{13.2}$$

The transfer function f is usually some continuous function mapping the real numbers into the interval $(-1, 1)$. Choices for f include $f(x) = (1 + \exp^{-x})^{-1}$ called the sigmoidal function, $f(x) = (2/\pi)\arctan(x)$, and $y = f(x) = \tanh(x)$ (see Section 10.3). These functions, except the sigmoidal, have the following properties: (1) they are monotonically increasing with $f(0) = 0$, $f(x) < 0$ for $x < 0$ and $f(x) > 0$ for $x > 0$; and (2) $\lim_{x\to\infty} f(x) = 1$ and $\lim_{x\to-\infty} f(x) = -1$. The sigmoidal function has the following properties: (1) $0 < f(x) < 1$, (2) $f(0) = 0.5$, and (3) $lim_{x\to\infty} = 1$ and $lim_{x\to-\infty} = 0$. We assume all the transfer functions within the neural net are the same function.

The output from each neuron in the second layer is sent, along the directed arcs shown in Figure 13.1, to the output neuron. Each arc connecting neurons in the second layer to the output neuron has a weight u_i. The process described above is repeated at the output neuron so the input is

$$t = z_1 u_1 + z_2 u_2 + z_3 u_3. \tag{13.3}$$

The output neuron adds a shift term φ to t and inputs this to transfer function f. The final output from the neural net is

$$y = f(t + \varphi). \tag{13.4}$$

Some, or all, of these shift terms can be zero.

It is not too difficult now to generalize these computations to a "m–n–p" neural net with m input neurons, n neurons in the second layer and p output neurons. We will abbreviate a neural net as NN so the input–output relationships in equations (13.1)–(13.4) are summarized as $y = NN(x)$, for $x = (x_1, x_2)$. For an m–n–p neural net we also write $y = NN(x)$ for $y = (y_1, \cdots, y_p)$ and $x = (x_1, \cdots, x_m)$.

We are going to use neural nets to approximate values of functions. Let $y = g(x)$ be a continuous function for $x \in [a, b]$. Then it is well-known that there is a 1–m–1 neural net NN so that $NN(x) \approx g(x)$ for all x in $[a,b]$. The symbol "$\approx$" means that $NN(x)$ is approximately equal to $g(x)$ for all x in $[a,b]$. This approximation can be made less than 0.0001 or even less than 10^{-8}. By the phrase "there is a 1–m–1 neural net" means there exists a 1–m–1 neural net with m neurons in the second layer, there are also weights w_{ij}, u_i and shift terms and there is a transfer function f so that the required approximation holds. Let us fix the transfer function f as the sigmoidal transfer function. So given $y = g(x)$ and $[a,b]$ we can find m neurons for the second layer and all the weights and shift terms so that $NN(x) \approx g(x)$ for all x in $[a,b]$.

Now let us assume we know the value of m and the transfer function f is sigmoidal. How do we find the values for all the weights and shift terms? The weights and shift terms are found using a training (learning) algorithm and some data (x and y values) on the function g. We are not going to present a training algorithm in this book. There are now many training algorithms available by downloading them from websites. In your search engine use "neural nets", "neural networks", etc. to locate training (learning) software to download. For the rest of this chapter we will assume that any needed training (finding the weights and shift terms) has already been completed. One can also use a genetic algorithm (Chapter 15) for training a neural net.

The approximation result generalizes to $y = g(x_1, \cdots, x_n)$ a continuous function for $x_i \in [a_i, b_i]$, $1 \leq i \leq n$. Now there is an n–m–1 neural net NN so that $NN(x_1, \cdots, x_n) \approx g(x_1, \cdots, x_n)$ for all $x_i \in [a_i, b_i]$, $1 \leq i \leq n$.

Let us now look at how we are going to design a neural net to: (1) approximate solutions to fuzzy equations (Chapter 5), and (2) approximate values of fuzzy functions.

We begin with a neural net to approximate solutions to

$$\overline{A} \cdot \overline{X} = \overline{C}, \tag{13.5}$$

for $\overline{A} = (a_1/a_2/a_3) < 0$, $\overline{C} = (c_1/c_2/c_3) > 0$ and unknown $\overline{X} \approx (x_1/x_2/x_3)$. The neural net will only approximate $\overline{X}_i$, the α-cut and interval arithmetic solution (see Chapter 5) to equation (13.5). The solution is obtained by fuzzifying the crisp solution. The crisp solution to $ax = c$, $a \neq 0$, is $x = c/a$. We fuzzify c/a by substituting α-cuts of $\overline{C}$ for c, α-cuts of $\overline{A}$ for a and simplify using interval arithmetic. We find that

$$\overline{X}_i = \frac{[c_1(\alpha), c_2(\alpha)]}{[a_1(\alpha), a_2(\alpha)]} = [\frac{c_2(\alpha)}{a_2(\alpha)}, \frac{c_1(\alpha)}{a_1(\alpha)}], \tag{13.6}$$

where $\overline{A}[\alpha] = [a_1(\alpha), a_2(\alpha)]$, $\overline{C}[\alpha] = [c_1(\alpha), c_2(\alpha)]$, because $\overline{A} < 0$.

Now we design the neural net. It will be a 2–m–1, the transfer function is sigmoidal , the identity transfer function in the input and output neurons, no shift terms in the input and output neurons and sign constraints on the weights. The neurons in the second layer have f and shift term θ_i and there are no sign constraints on the shift terms. By sign constraints on the weights we mean some must be negative and the others must be positive. There are training algorithms for neural nets with sign constraints on the weights.

Assume $\overline{A}$ is a triangular fuzzy number in the interval $I_a = [a_l, a_u]$, $a_u < 0$ and $\overline{C}$ is another triangular fuzzy number in the interval $I_c = [c_l, c_u]$, $c_l > 0$. We now demand that all $w_{ij} < 0$ and all $u_i > 0$. The reasons for these sign constraints will become clear as we work through the whole problem. Train the NN, with two inputs $x_1 = c \in I_c$ and $x_2 = a \in I_a$, to approximate c/a. That is, we have a continuous function $y = g(a, c) = c/a$ for $a \in I_a$ and $c \in I_c$, so there is a 2–m–1 neural net, as described above, with $w_{ij} < 0$, $u_i > 0$, so that $NN(a, c) \approx c/a$ for all $a \in I_a$ and all $c \in I_c$.

Once trained, input $x_1 = [c_1(\alpha), c_2(\alpha)]$ and $x_2 = [a_1(\alpha), a_2(\alpha)]$. We now argue that the output y is an interval $[y_1(\alpha), y_2(\alpha)]$ and

$$[y_1(\alpha), y_2(\alpha)] \approx [\frac{c_2(\alpha)}{a_2(\alpha)}, \frac{c_1(\alpha)}{a_1(\alpha)}], \tag{13.7}$$

for all α. That is, the output approximates α-cuts of $\overline{X}_i$. We stress that now the neural net processes intervals and not just real numbers. Also, within the neural net it does interval arithmetic. We now go through all the details on how you compute the output.

The input to the jth neuron in the second layer is

$$w_{1j}[c_1(\alpha), c_2(\alpha)] + w_{2j}[a_1(\alpha), a_2(\alpha)], \tag{13.8}$$

which equals

$$[c_2(\alpha)w_{1j} + a_2(\alpha)w_{2j}, c_1(\alpha)w_{1j} + a_1(\alpha)w_{2j}], \tag{13.9}$$

because $w_{1j} < 0$ and $w_{2j} < 0$. Add the shift term θ_i (a real number, not an interval) and input to transfer function f. The output from the jth neuron in the second layer is an interval $[z_{j1}, z_{j2}]$ where

$$z_{j1} = f(c_2(\alpha)w_{1j} + a_2(\alpha)w_{2j} + \theta_j), \tag{13.10}$$

and

$$z_{j2} = f(c_1(\alpha)w_{1j} + a_1(\alpha)w_{2j} + \theta_j), \tag{13.11}$$

because f is monotonically increasing. Now the output equals the input in the output neuron so if $y = [y_1(\alpha), y_2(\alpha)]$ is the interval output we see that

$$y_1(\alpha) = u_1 z_{11} + \cdots + u_m z_{m1}, \tag{13.12}$$

and

$$y_2(\alpha) = u_1 z_{12} + \cdots + u_m z_{m2}, \tag{13.13}$$

because all the $u_i > 0$. We claim that $y_1(\alpha) \approx c_2(\alpha)/a_2(\alpha)$ and $y_2(\alpha) \approx c_1(\alpha)/a_1(\alpha)$ for all $\alpha \in [0, 1]$.

To see this we need to go back to the original trained neural net and $y \approx c/a$. If the inputs are $x_1 = c$ and $x_2 = a$, then

$$y = \sum_{j=1}^{m} u_j f(cw_{1j} + aw_{2j} + \theta_j) \approx c/a. \tag{13.14}$$

But from the interval neural net ($w_{ij} < 0, u_i > 0$) we have

$$y_1(\alpha) = \sum_{j=1}^{m} u_j f(c_2(\alpha)w_{1j} + a_2(\alpha)w_{2j} + \theta_j), \tag{13.15}$$

and

$$y_2(\alpha) = \sum_{j=1}^{m} u_j f(c_1(\alpha)w_{1j} + a_1(\alpha)w_{2j} + \theta_j). \tag{13.16}$$

Substitute $c_2(\alpha)$ for c, $a_2(\alpha)$ for a in equation (13.14) and you get equation (13.15). Hence $y_1(\alpha) \approx c_2(\alpha)/a_2(\alpha)$. Also substitute $c_1(\alpha)$ for c and $a_1(\alpha)$ for a in equation (13.14) and you get equation (13.16). It follows that $y_2(\alpha) \approx c_1(\alpha)/a_1(\alpha)$. Now do you see why we needed all $w_{ij} < 0$ and all $u_i > 0$?

Next, let us design a neural net with sign constraints to approximate the values of the fuzzy function

$$\overline{Y} = \overline{F}(\overline{X}) = \frac{\overline{A} - \overline{X}}{\overline{B} \cdot \overline{X} + 1}, \tag{13.17}$$

for triangular fuzzy numbers $\overline{A}, \overline{B}, \overline{X}$ and $\overline{A} > 0$, $\overline{B} > 0$ and $\overline{X} > 0$. Assume that $\overline{A}(\overline{B}, \overline{X})$ are in intervals $I_a(I_b, I_x)$.

We train a 3–m–1 neural net, with inputs $x_1 = a$, $x_2 = b$ and $x_3 = x$, with sign constraints $w_{2j} < 0$, $w_{3j} < 0$ and all other weights positive, so that $NN(a, b, x) \approx (a - x)/(bx + 1)$, for all $a \in I_a$, $b \in I_b$ and $x \in I_x$.

Now make the trained neural net process intervals. Input $\overline{A}[\alpha] = x_1$, $\overline{B}[\alpha] = x_2$, $\overline{X}[\alpha] = x_3$ and let the interval output be $y = [y_1(\alpha), y_2(\alpha)]$. We claim that

$$y_1(\alpha) \approx \frac{a_1(\alpha) - x_2(\alpha)}{b_2(\alpha)x_2(\alpha) + 1}, \tag{13.18}$$

and

$$y_2(\alpha) \approx \frac{a_2(\alpha) - x_1(\alpha)}{b_1(\alpha)x_1(\alpha) + 1}. \tag{13.19}$$

If we substitute α-cuts of $\overline{A}, \overline{B}, \overline{X}$ into equation (13.17) and simplify using interval arithmetic, then we get an interval $[L(\alpha), R(\alpha)]$. Hence, $L(\alpha)$ is the right side of equation (13.18) and $R(\alpha)$ is the right side of equation (13.19).

So, the interval output from the trained neural net , which processes α-cuts of $\overline{A}, \overline{B}, \overline{X}$, approximate $[L(\alpha), F(\alpha)]$, the α-cut and interval arithmetic evaluation of $\overline{F}(\overline{X})$.

13.2.1 Exercises

You are asked to design a neural net for a certain problem. The neural net will be k–m–1. The identity transfer function, and no shift terms, are in all input neurons and also in the output neuron. Assume the sigmoidal transfer function for all neurons in the second layer. There will be sign constraints on the w_{ij} and the u_i but no sign constraints on the shift terms θ_i.

To "design" the neural net determine the value of k and the sign constraints on the w_{ij} and u_i. Also tell all inputs to the interval neural net: $x_1 =?, \cdots, x_k =?$. Assume all fuzzy sets are triangular fuzzy numbers.

1. For $\overline{X}_i$ being a solution to $\overline{A} \cdot \overline{X} = \overline{C}$ where:

   a. $\overline{A} > 0, \overline{C} > 0$,
   b. $\overline{A} > 0, \overline{C} < 0$,
   c. $\overline{A} < 0, \overline{C} < 0$.

2. For $\overline{X}_i$ being a solution to $\overline{A} \cdot \overline{X} + \overline{B} = \overline{C}$ when:

   a. $\overline{A} > 0, \overline{B} > 0, \overline{C} > 0$;
   b. $\overline{A} > 0, \overline{B} > 0, \overline{C} < 0$;
   c. $\overline{A} > 0, \overline{B} < 0, \overline{C} > 0$.
   d. $\overline{A} > 0, \overline{B} < 0, \overline{C} > 0$;
   e. $\overline{A} < 0, \overline{B} > 0, \overline{C} > 0$;
   f. $\overline{A} < 0, \overline{B} > 0, \overline{C} < 0$;
   g. $\overline{A} < 0, \overline{B} < 0, \overline{C} > 0$; and
   h. $\overline{A} < 0, \overline{B} < 0, \overline{C} < 0$.

3. For $\overline{Y} = \overline{F}(\overline{X})$ when $\overline{F}(\overline{X}) = \overline{A} \cdot \overline{X}^2 + \overline{B} \cdot \overline{X} + \overline{C}$, for:

   a. $\overline{A} > 0, \overline{B} > 0, \overline{C} > 0$ but $\overline{X} < 0$;
   b. $\overline{A} > 0, \overline{B} < 0, \overline{C} < 0$ and $\overline{X} > 0$;
   c. $\overline{A} > 0, \overline{B} < 0, \overline{C} < 0$ and $\overline{X} < 0$; and
   d. $\overline{A} > 0, \overline{B} < 0, \overline{C} > 0$ and $\overline{X} > 0$.

4. For $\overline{Y} = \overline{F}(\overline{X}) = \overline{A} \exp(\overline{B} \cdot \overline{X} + \overline{C})$ for:

   a. All positive fuzzy numbers;
   b. $\overline{A} > 0, \overline{B} > 0, \overline{X} > 0$ but $\overline{C} < 0$;
   c. $\overline{A} > 0, \overline{B} < 0, \overline{X} > 0, \overline{C} > 0$; and
   d. $\overline{A} > 0, \overline{B} < 0, \overline{X} > 0, \overline{C} < 0$.

5. For $\overline{Y} = \overline{F}(\overline{X}) = \overline{A} \ln(\overline{B} \cdot \overline{X} + \overline{C})$ for (in all cases $\overline{B} \cdot \overline{X} + \overline{C} > 0$):

a. All positive fuzzy numbers;

b. $\overline{A} > 0, \overline{B} > 0, \overline{X} > 0, \overline{C} < 0$;

c. $\overline{A} > 0, \overline{B} < 0, \overline{X} < 0, \overline{C} > 0$; and

d. $\overline{A} < 0, \overline{B} < 0, \overline{X} < 0, \overline{C} > 0$.

6. For $\overline{Y}$ in equation (13.17) for:

a. $\overline{A} < 0$, $\overline{B} > 0$, $\overline{X} > 0$;

b. $\overline{A} > 0$, $\overline{B} < 0$, $\overline{X} > 0$; and

c. $\overline{A} < 0$, $\overline{B} < 0$, $\overline{X} < 0$.

7. Find, and download, training software for layered, feedforward, neural nets allowing for sign constraints on the weights. Then do the training for the example $\overline{A} \cdot \overline{X} = \overline{C}$, $\overline{A} < 0$ the rest positive, in the text. Find the α-cuts for $\overline{X}$ using the trained neural net and compare to the exact values given in equation (13.6).

13.3 Fuzzy Neural Nets

In Figure 13.1 we get a fuzzy neural net when some of the x_i, w_{ij}, u_i and shift terms are continuous fuzzy numbers. We have a Type I fuzzy neural net when the only thing fuzzy is the input. Type II when everything can be fuzzy except the inputs. If the inputs, weights and shift terms are all fuzzy we call it a Type III fuzzy neural net .

Let us look at a Type III fuzzy neural net and see how the output $\overline{Y}$ is computed from the inputs $\overline{X}_1$ and $\overline{X}_2$. The basic computations are the same as shown in equations (13.1)–(13.4). Assume a 2–3–1 structure as in Figure 13.1. First compute $\overline{X}_i\overline{W}_{ij}$ for $i = 1, 2$ and $j = 1, 2, 3$. Then add to get $\overline{X}_1\overline{W}_{1j} + \overline{X}_2\overline{W}_{2j} + \overline{\Theta}_j$ for $j = 1, 2, 3$. Put this result into the transfer function f producing output $\overline{Z}_j$ for $j = 1, 2, 3$. The $\overline{Z}_j$ are easily found because f will be monotonically increasing. Multiply again $\overline{Z}_j\overline{U}_j$ for $j = 1, 2, 3$ and add $\overline{W} = \overline{Z}_1\overline{U}_1 + \cdots + \overline{Z}_3\overline{U}_3 + \overline{\varphi}$. Then $\overline{Y} = f(\overline{W})$ is the final output. We went through this step-by-step to show that the extension principle method and the α-cut and interval arithmetic procedure will give the same result for $\overline{Y}$. We mentioned in Section 4.3.3 that for elementary fuzzy arithmetic both methods produce the same result. That is just what we did, except for using a transfer function, to compute $\overline{Y}$ the output. Both procedures give the same result for f because the transfer function is monotonically increasing. So, we will employ the α-cut and interval arithmetic method within a fuzzy neural net to obtain the output.

What we would do within a computer to find the output $\overline{Y}$ is to input α-cuts, say $\alpha = 0, 0.1, \cdots, 1$, of the fuzzy inputs $\overline{X}_1, \overline{X}_2$, do interval arithmetic within the fuzzy neural net, then "stack up" the interval outputs to approximate $\overline{Y}$. A Type I fuzzy neural net becomes a neural net of Section 13.2 processing interval input.

Let FNN denote a fuzzy neural net so that: (1) $FNN(\overline{X}) = \overline{Y}$, $\overline{X} = (\overline{X}_1, \overline{X}_2)$, for Type I; (2) $FNN(x) = \overline{Y}$, $x = (x_1, x_2)$, for Type II; and (3) $FNN(\overline{X}) = \overline{Y}$ for Type III. $\overline{X}$ and x are vectors but also $\overline{Y}$ could be a vector. Type I and Type III are monotone increasing fuzzy functions. Let $\overline{X}_a = (\overline{X}_{a1}, \overline{X}_{a2})$ and $\overline{X}_b = (\overline{X}_{b1}, \overline{X}_{b2})$ be two possible input vectors. Monotone increasing means that if $\overline{X}_{ai} \leq \overline{X}_{bi}$ for $i = 1, 2$, then $\overline{Y}_a = FNN(\overline{X}_a) \leq FNN(\overline{X}_b) = \overline{Y}_b$. The fact that Type I and Type III fuzzy neural nets are monotone increasing is important because if we want to use a fuzzy neural net to approximate the values of some fuzzy function $\overline{Y} = \overline{F}(\overline{X})$, then $\overline{F}$ must also be monotone increasing.

We will not need to discuss training of fuzzy neural nets in this section. However, one can use genetic algorithms (Chapter15) to do the training (also for crisp neural nets).

What we want to do now is build hybrid fuzzy neural nets for fuzzy functions. A FNN is called a hybrid FNN when we use for transfer functions some other continuous operation, not one of those discussed in Section 13.2, within a neuron. The transfer function could be multiplication, division, etc.

For example, in Figure 13.1 for a hybrid Type III fuzzy neural net let the output from the second neuron in the second layer be

$$\overline{Z}_2 = (\overline{X}_1 \overline{W}_{12})(\overline{X}_2 \overline{W}_{22}). \tag{13.20}$$

We multiplied the incoming weighted signals (assume the fuzzy shift term is zero). The transfer function was multiplication. The transfer function could have been division, or any other continuous operation. In a hybrid FNN there will be no restrictions, except that it is continuous, on how we might combine the incoming weighted signals.

These hybrid FNNs will be built to do a specific job. There will be no training (learning) algorithm needed. Now let HFNN denote a hybrid fuzzy neural net. If we construct a HFNN for a fuzzy function $\overline{Y} = \overline{F}(\overline{X})$, then $HFNN(\overline{X}) = \overline{F}(\overline{X})$ for all $\overline{X}$, this is an exact result and not an approximation.

Example 13.3.1

Build a HFNN for

$$\overline{Y} = \overline{F}(\overline{X}) = \frac{2\overline{X} + 3}{\overline{X} + 1}, \tag{13.21}$$

for $\overline{X} > 0$ a continuous fuzzy number. The HFNN is shown in Figure 13.2. This is a Type I HFNN. If no transfer function "f" is included within a neuron, then it is to be the identity ($i(x) = x$) function.

The input to neuron #1 in the second layer is $2\overline{X}$, it has the identity transfer function and the shift term is 3, so its output is $2\overline{X} + 3$. For neuron #2 in the second layer, its input equals $1\overline{X}$, identity transfer, shift term equal to one, and then its output is $\overline{X} + 1$. The two inputs to the output neuron are $2\overline{X} + 3$ and $\overline{X} + 1$, the output neuron performs division with no shift term, so the final output is $(2\overline{X} + 3)/(\overline{X} + 1)$ the value of $\overline{F}$.

Example 13.3.2

A Type II HFNN for the fuzzy function

$$\overline{Y} = \overline{F}(t) = (\overline{A}e^{-t})(\overline{B}t + \overline{C}), \tag{13.22}$$

is in Figure 13.3. The function in neuron #1 in the second layer is $f(t) = e^{-t}$, which is its output (zero shift term). The input to neuron #2 in the second layer is $\overline{B}t$, it has the identity transfer function and shift term $\overline{C}$, so its output is $\overline{B}t + \overline{C}$. The two inputs to the output neuron are $\overline{A}e^{-t}$ and $\overline{B}t + \overline{C}$ and the output neuron performs multiplication, so its output equals $\overline{F}(t)$.

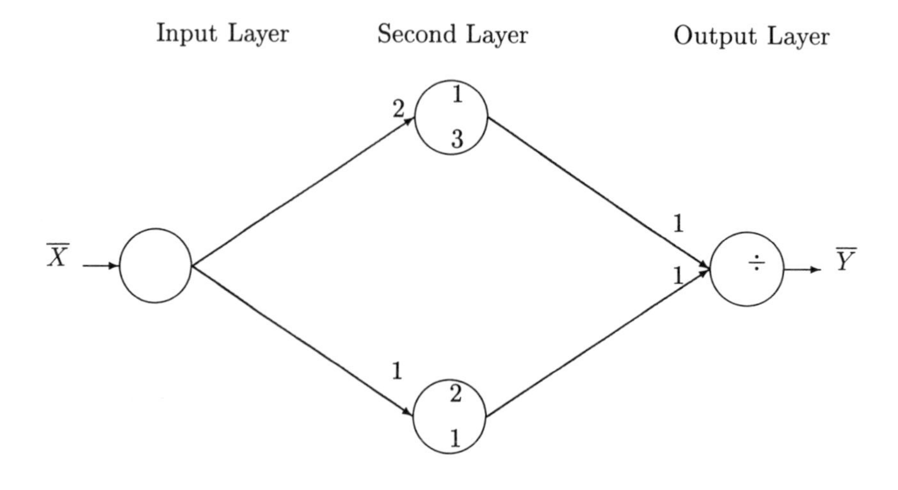

Figure 13.2: Hybrid Fuzzy Neural Net for Example 13.3.1

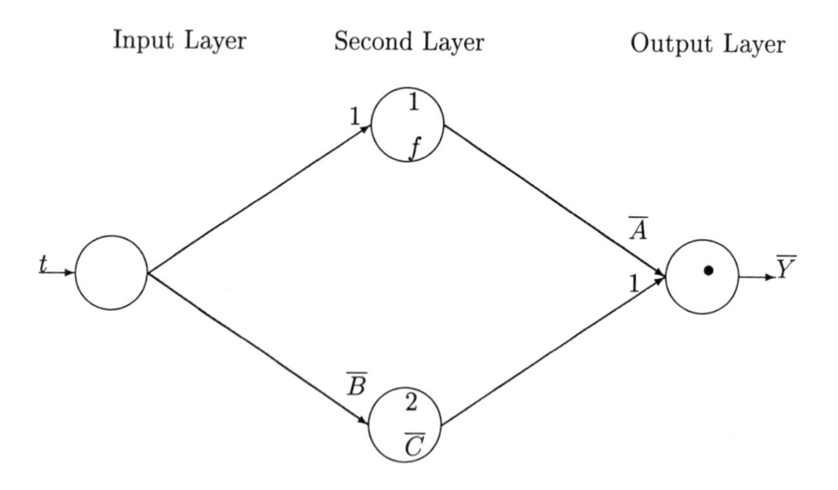

Figure 13.3: Hybrid Fuzzy Neural Net for Example 13.3.2

Example 13.3.3

A Type III HFNN for

$$\overline{Z} = \overline{F}(\overline{X}, \overline{Y}) = \overline{A} \cdot \overline{X}^2 + \overline{B} \cdot \overline{X} \cdot \overline{Y} + \overline{C} \cdot \overline{Y}^2, \tag{13.23}$$

is shown in Figure 13.4. It is a 2 - 3 - 3 - 1 HFNN, the only time within the text of this book that we shall use more than three layers. All neurons, except in the second layer, have the identity transfer function, with no shift terms. In the second layer $f_1(x) = f_3(x) = x^2$ but $f_2(x, y) = xy$. All the shift terms in the second layer are zero. So the output from the second layer is: (1) $\overline{X}^2$ for neuron #1; (2) $\overline{X} \cdot \overline{Y}$ from neuron #2; and (3) neuron #3 gives $\overline{Y}^2$. The outputs from the third layer are: (1) $\overline{A} \cdot \overline{X}$ from neuron #1; (2) neuron #2 produces $\overline{B} \cdot \overline{X} \cdot \overline{Y}$; and (3) $\overline{C} \cdot \overline{Y}^2$ from neuron #3. The output neuron adds all its inputs, no shift term, so the final output equals $\overline{F}(\overline{X}, \overline{Y})$.

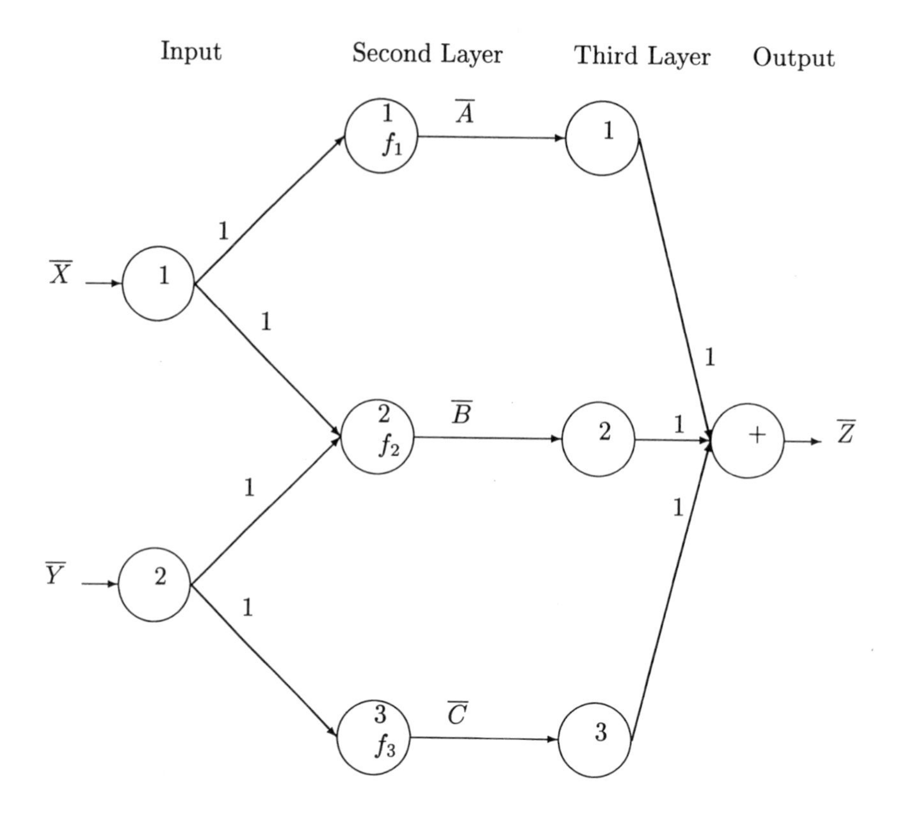

Figure 13.4: Hybrid Fuzzy Neural Net for Example 13.3.3

13.3.1 Exercises

1. Show that a Type I FNN is a monotone increasing fuzzy function.

2. Show that a Type III FNN is a monotone increasing fuzzy function.

In the following problems you are asked to design a HFNN to compute the values of a fuzzy function. Draw a network diagram, indicate all weights and shift terms and define all transfer functions. You may have to use more than three layers. Indicate all inputs to the HFNN. Also state if it is to be a Type I, or II , or III HFNN. The $\overline{A}, \cdots, \overline{E}$ are all constants and $\overline{X}$, $\overline{Y}$, $\overline{Z}$, and $\overline{W}$ and t are the variables.

3. For the fuzzy linear functions:

   a. $\overline{Y} = \overline{A}t + \overline{B}$,
   b. $\overline{Y} = 6\overline{X} + 10$, and
   c. $\overline{Y} = \overline{A} \cdot \overline{X} + \overline{B}$.

4. For the fuzzy quadratic functions:

   a. $\overline{Y} = 6\overline{X}^2 - 2\overline{X} + 10$,
   b. $\overline{Y} = \overline{A}t^2 + \overline{B}t + \overline{C}$, and
   c. $\overline{Y} = \overline{A} \cdot \overline{X}^2 + \overline{B} \cdot \overline{X} + \overline{C}$.

5. For the fuzzy rational functions:

   a. $\overline{Y} = \frac{\overline{A}t + \overline{B}}{\overline{C}t + \overline{D}}$, and
   b. $\overline{Y} = \frac{\overline{A} \cdot \overline{X} + \overline{B}}{\overline{C} \cdot \overline{X} + \overline{D}}$.

6. For the fuzzy exponential functions:

   a. $\overline{Y} = 6\exp(-2\overline{X} + 4)$,
   b. $\overline{Y} = \overline{A}\exp(\overline{B}t + \overline{C})$, and
   c. $\overline{Y} = \overline{A}\exp(\overline{B} \cdot \overline{X} + \overline{C})$.

7. For the fuzzy log functions:

   a. $\overline{Y} = 6\ln(2\overline{X} + 4)$,
   b. $\overline{Y} = \overline{A}\ln(\overline{B}t + \overline{C})$, and
   c. $\overline{Y} = \overline{A}\ln(\overline{B} \cdot \overline{X} + \overline{C})$.

8. For the fuzzy sine functions (input to sine is in the interval $[-\pi/2, \pi/2]$):

   a. $\overline{Y} = 2\sin(\pi\overline{X} + 1)$,
   b. $\overline{Y} = \overline{A}\sin(\overline{B}t + \overline{C})$, and

c. $\overline{Y} = \overline{A}\sin(\overline{B} \cdot \overline{X} + \overline{C})$.

9. For the fuzzy tangent functions:

   a. $\overline{Y} = -4\tan((\pi/2)\overline{X} - 1)$,
   b. $\overline{Y} = \overline{A}\tan(\overline{B}t + \overline{C})$, and
   c. $\overline{Y} = \overline{A}\tan(\overline{B} \cdot \overline{X} + \overline{C})$.

10. For the fuzzy hyperbolic sine functions:

    a. $\overline{Y} = 26\sinh(7\overline{X} + 13)$,
    b. $\overline{Y} = \overline{A}\sinh(\overline{B}t + \overline{C})$, and
    c. $\overline{Y} = \overline{A}\sinh(\overline{B} \cdot \overline{X} + \overline{C})$.

11. For the fuzzy hyperbolic tangent functions:

    a. $\overline{Y} = -4\tanh(3\overline{X} + 2)$,
    b. $\overline{Y} = \overline{A}\tanh(\overline{B}t + \overline{C})$, and
    c. $\overline{Y} = \overline{A}\tanh(\overline{B} \cdot \overline{X} + \overline{C})$.

12. $\overline{Z} = \overline{A} \cdot \overline{X}^2 + \overline{B} \cdot \overline{X} \cdot \overline{Y} + \overline{C} \cdot \overline{Y}^2 + \overline{D} \cdot \overline{X} + \overline{E} \cdot \overline{Y}$.

13. $\overline{W} = \overline{X} - \sqrt{\overline{Y}^2 + \overline{Z}^2}$.

14. $\overline{W} = (\exp(3\overline{X} + 4\overline{Y}))(\cos(5\overline{Z}))$. Assume the input to cosine is in the interval $[0, \pi]$.

Chapter 14

Approximate Reasoning

14.1 Introduction

A method of processing information (data) through fuzzy rules is called approximate reasoning. If we have only one fuzzy rule like "if size is big, then speed is slow", and we are given a (fuzzy) value for size, then approximate reasoning gives us a method of computing a conclusion about speed. The terms "big", "slow" and the data for "size" are all represented as fuzzy sets. The single rule case is discussed in the next section and multiple fuzzy rules are studied in the third section. Also in the third section of this chapter we look at two methods of evaluating a block of fuzzy rules: (1) FITA, or first infer and then aggregate; and (2) FATI, or first aggregate and then infer.

In computer applications we usually discretize all the continuous fuzzy numbers and this, for multiple fuzzy rules, is explained in the fourth section. There are situations of fuzzy rules where approximate reasoning can not be used. This happens when we can not assign fuzzy numbers to all the terms in a fuzzy rule and examples of this , together with fuzzy rule evaluation for this type of fuzzy rule, is in the last section, section five.

14.2 Approximate Reasoning

Approximate reasoning is concerned with evaluating a fuzzy rule like

$$if\ x\ is\ \overline{A},\ then\ y\ is\ \overline{B}. \tag{14.1}$$

In this rule we have two universal sets X and Y, $x(y)$ is a variable taking its values in X(Y) and $\overline{A}$(or $\overline{B}$) is a fuzzy subset of X(or Y). Later on we will consider more than one rule.

Where do these fuzzy sets $\overline{A}$ and $\overline{B}$ come from? Usually these fuzzy sets come from the values of linguistic variables. Consider two linguistic variables Size and Speed. Size has members *tiny, small, medium, large, very large* and

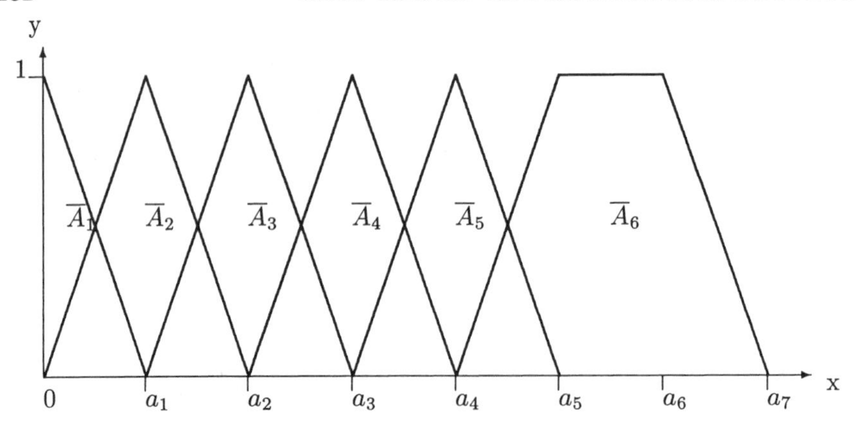

Figure 14.1: Fuzzy Numbers for Linguistic Variable Size

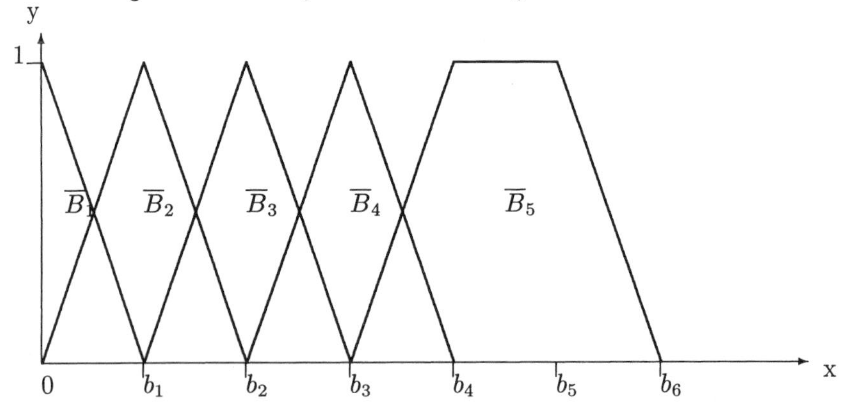

Figure 14.2: Fuzzy Numbers for Linguistic Variable Speed

huge. The members of Size are defined by fuzzy numbers and an example of this is shown in Figure 14.1.

We have $\overline{A}_1$ for *tiny*, $\overline{A}_2$ for *small*, $\cdots$, and $\overline{A}_6$ for *huge*. We used triangular fuzzy numbers for $\overline{A}_2, \cdots, \overline{A}_5$, a trapezoidal fuzzy number $\overline{A}_6$ for *huge* and half a triangular fuzzy number $\overline{A}_1$ (since $x \geq 0$) for *tiny*.

Speed has members *very slow, slow, medium, fast* and *very fast*, all defined by fuzzy numbers an example of which is in Figure 14.2.

We use $\overline{B}_1$ for *very slow*, $\cdots$, and $\overline{B}_5$ for *very fast*.

With this new information the rule in equation (14.1) might now read

$$if\ x\ is\ very\ large,\ then\ y\ is\ slow. \tag{14.2}$$

For *very large* we substitute $\overline{A}_5$ and for *slow* we would use $\overline{B}_2$.

Given some data $x = \overline{A}'$, a non-negative continuous fuzzy number representing the size of some object, what should be the conclusion $y = \overline{B}'$ about its speed? First we usually choose some implication operator I from classical logic. Implication operators I were discussed in Chapter 2 (Section 2.2). I is to model the implication

$$(x \text{ } is \text{ } \overline{A}) \rightarrow (y \text{ } is \text{ } \overline{B}), \tag{14.3}$$

in the rule. $I(a, b)$ will be a number in $[0, 1]$ for all $a, b \in [0, 1]$. We fuzzify I by replacing "a" by $\overline{A}(x)$ and substituting $\overline{B}(y)$ for "b" producing a fuzzy relation (Chapter 7) $\overline{R}$ on $X \times Y$ defined by

$$\overline{R}(x, y) = I[\overline{A}(x), \overline{B}(y)]. \tag{14.4}$$

Example 14.2.1

Choose the Łukasiewicz implication operator $I(a, b) = \min(1, 1 - a + b)$. So $\overline{R}(x, y) = \min(1, 1 - \overline{A}(x) + \overline{B}(y))$. For continuous fuzzy numbers $\overline{A}$ and $\overline{B}$, so X and Y are the sets of real numbers, $z = \overline{R}(x, y)$ will be a surface in $\mathbf{R}^3$.

Given the data $x = \overline{A'}$ we compute the conclusion $y = \overline{B'}$ using the compositional rule of inference. We discussed the composition of fuzzy relations in Section 7.2 of Chapter 7. Symbolically it is written as $\overline{B'} = \overline{A'} \circ \overline{R}$. From the composition of fuzzy relations we obtain the membership function for $\overline{B'}$ as follows

$$\overline{B'}(y) = \sup_x \{\min\{\overline{A'}(x), \overline{R}(x, y)\}\}, \tag{14.5}$$

for each $y \in Y$. Notice that we used t-norm $T_m = \min$ in equation (14.5). We may also use T_b, T_p or T^*.

We are only using one rule in equation (14.5). But with only one rule the whole process is a mapping (a fuzzy function) from fuzzy subsets $\overline{A'}$ of X into fuzzy subsets $\overline{B'}$ of Y. Let us call this fuzzy function $\overline{AR}$, for approximate reasoning, so that

$$\overline{B'} = \overline{AR}(\overline{A'}), \tag{14.6}$$

for $\overline{A'}$ fuzzy subset of X and $\overline{B'}$ fuzzy subset of Y. $\overline{AR}$ depends on I, the implication operator used, and on the t-norm employed in equation (14.5).

In general it is difficult to evaluate equation (14.5) for continuous fuzzy numbers, but if the input is crisp, we may more easily find the conclusion $\overline{B'}$.

Example 14.2.2

Consider crisp input

$$\overline{A'}(x) = \begin{cases} 1, & x = x_0, \\ 0, & \text{otherwise.} \end{cases} \tag{14.7}$$

Then we see, using the Lukasiewicz implication operator from Example 14.2.1, that

$$\overline{B'}(y) = \min(1, 1 - \overline{A}(x_0) + \overline{B}(y)), \tag{14.8}$$

all $y \in Y$. Now $\overline{B'}(y)$ depends on the value of $\overline{A}(x_0)$. If $\overline{A}(x_0) = 1$, then $\overline{B'} = \overline{B}$. If $\overline{A}(x_0) = 0$, then $\overline{B'}(y) = 1$ for all y. We obtain complete uncertainty in our conclusion (membership function identically one) for $\overline{A}(x_0) = 0$.

Example 14.2.3

There is a popular method of determining a conclusion $\overline{B'}$ from an input $\overline{A'}$ not depending on an implication operator from classical logic. For the fuzzy rule in equation (14.1) the fuzzy relation is simply

$$\overline{R}(x, y) = \min(\overline{A}(x), \overline{B}(y)). \tag{14.9}$$

Then we find the conclusion $\overline{B'}$ as in equation (14.5). If we have crisp input as in Example 14.2.2, then

$$\overline{B'}(y) = \min(\overline{A}(x_0), \overline{B}(y)). \tag{14.10}$$

If $\overline{A}(x_0) = 0.6$, then

$$\overline{B'}(y) = \begin{cases} \overline{B}(y), & \overline{B}(y) \leq 0.6, \\ 0.6, & \overline{B}(y) > 0.6. \end{cases} \tag{14.11}$$

The "top" of $\overline{B}(y)$ was cut off at height 0.6 to determine the conclusion.

There are two basic and important properties desired of approximate reasoning. The first one is consistency, or if the input data $x = \overline{A'}$ matches the antecednt of the fuzzy rule [the $\overline{A}$ in equation (14.1)], then the conclusion should match the conclusion of the fuzzy rule [the $\overline{B}$ of equation (14.1)]. That is, if $x = \overline{A'} = \overline{A}$, then $y = \overline{B'} = \overline{B}$. In the approximate reasoning function notation

$$\overline{B} = \overline{AR}(\overline{A}), \tag{14.12}$$

for consistency.

For consistency from equation (14.5) we need

$$\overline{B}(y) = \sup_x\{\min\{\overline{A}(x), I(\overline{A}(x), \overline{B}(y))\}, \tag{14.13}$$

using implication operator I and the t-norm min.

Example 14.2.4

Let $\overline{A} = (1/2/3)$ and $\overline{B} = (10/12/14)$ in equation (14.1) and keep I as the Łukasiewicz implication operator in Example 14.2.2. To show that this method of approximate reasoning is not consistent all we need to do is find some value of y, say y_0, so that $\overline{B'}(y_0) \neq \overline{B}(y_0)$, where we find $\overline{B'}$ from equation (14.5) using $\overline{A'} = \overline{A}$. Choose $y_0 = 11$. Then $\overline{B}(11) = 0.5$. Equation (14.5) gives

$$\overline{B'}(11) = \max_x\{\min\{\overline{A}(x), \min(1, 1.5 - \overline{A}(x))\}\}, \tag{14.14}$$

which equals 0.75. This method of approximate reasoning is not consistent.

Example 14.2.5

Keep the same $\overline{A}$ and $\overline{B}$ as in Example 14.2.4 but change I to

$$I(a,b) = \begin{cases} 1, & a \leq b \\ b, & a > b \end{cases} \tag{14.15}$$

called the Gödel implication operator. Form $\overline{R}(x,y) = I(\overline{A}(x), \overline{B}(y))$ and evaluate equation (14.5) using $\overline{A'} = \overline{A}$. Now we have consistency $\overline{B} = \overline{AR}(\overline{A})$.

Consistency comes from the modus ponens in classical logic: if $a \rightarrow b$ and given "a", then "b". The classical modus tollens is: if $a \rightarrow b$ and given $\bar{b}$ (not b), then conclude $\bar{a}$ (not a). This brings us to the second desired property of approximate reasoning: given $y = \overline{B}^c$ in the fuzzy rule in equation (14.1), then the conclusion should be $x = \overline{A}^c$. What this means is input $\overline{B}^c$ for $\overline{A'}$ and get $\overline{A}^c$ for $\overline{B'}$. Or

$$\overline{A}^c = \overline{AR}(\overline{B}^c). \tag{14.16}$$

Example 14.2.6

This continues Example 14.2.5, same $\overline{A}, \overline{B}$ and I. We wish to check to see if

$$1 - \overline{A}(x) = \max_y\{T_b\{1 - \overline{B}(y), \overline{R}(x,y)\}\}, \tag{14.17}$$

for all x. We may argue (see the exercises) that this equation does hold so that $\overline{A}^c = \overline{AR}(\overline{B}^c)$ is true for this implication operator I and t-norm T_b.

The fuzzy inference scheme given by the rule "if x is $\overline{A}$, then y is $\overline{B}$, given x is $\overline{A'}$, conclude y is $\overline{B'}$" is called the generalized modus ponens. The fuzzy inference scheme given by “if x is $\overline{A}$, then y is $\overline{B}$, given y is $\overline{B'}$, conclude x is $\overline{A'}$ ” is called the generalized modus tollens.

14.2.1 Exercises

1. Let R be a crisp relation on the set of real numbers. That is, R is a subset of $\mathbf{R} \times \mathbf{R}$ and we write $R(x,y) = 1$ if $(x,y) \in R$ and $R(x,y) = 0$ otherwise. Relation R models the implication

$$if\ x\ is\ in\ A,\ then\ y\ is\ in\ B,$$

for A and B crisp subsets of $\mathbf{R}$.

Let

$$B_1 = \{y \in \mathbf{R} | R(x,y) = 1,\ x \in A\},$$

and

$$B_2 = \max_x \{\min\{A(x), R(x,y)\}\}.$$

We write (as in Chapter 3) $A(x)$ for the characteristic function of A ($A(x) = 1$ if and only if $x \in A$). Show that $B_1 = B_2$. The expression for B_2 is the crisp version of equation (14.5). Does this result depend on the fact that the universal set was the set of real numbers? That is, could we get the same conclusion if R was a crisp relation on arbitrary set X?

2. Let I be the Łukasiewicz implication operator of Example 14.2.1. Using t-norm min and equation (14.5) show that if $\overline{A'} \cap \overline{A} = \overline{\phi}$ (with t-norm min for intersection), then $\overline{B'}(y) = 1$ for all y in Y (complete uncertainty in the conclusion).

3. Use $\overline{A}$ and $\overline{B}$ from Example 14.2.4 and the Łukasiewicz implication operator I. Suppose $\overline{A'} = (2/3/4)$. Show that $\overline{B'}(y) \geq 0.5$ for all y in Y. Is this a desirable or undesirable result of approximate reasoning?

4. Use the $\overline{R}$ from Example 14.2.3. Show that if $\overline{A'} \cap \overline{A} = \overline{\phi}$ (with min for intersection), then equation (14.5) gives $\overline{B'}(y) = 0$ for all y in Y. Comment on this result as being desirable or undesirable for approximate reasoning.

5. Show that in Example 14.2.4 that $\overline{B'}(11)$ does equal 0.75.

6. Redo Example 14.2.4 using t-norm T_b for min in equation (14.5).

7. Show that in Example 14.2.5 that you do get $\overline{B'} = \overline{B}$.

8. Redo Example 14.2.5 using t-norm T^* for min in equation (14.5).

9. In Example 14.2.6 show that $\overline{AR}(\overline{B^c}) = \overline{A^c}$. Would this result also be true if we used t-norm T_m?

10. Redo Example 14.2.5 with the data $\overline{A} = (10/12/14)$ and $\overline{B} = (1/2/3)$. Do we still get $\overline{B} = \overline{AR}(\overline{A})$? Do you think this result will hold for all $\overline{A}$ fuzzy subsets of X and all $\overline{B}$ fuzzy subsets of Y?

11. Consider the Gaines-Rescher implication operator which is: $I(a, b) = 1$ if $a \leq b$ and $I(a, b) = 0$ otherwise. Using the $\overline{A}$ and $\overline{B}$ from Example 14.2.4 and equation (14.5) do we obtain $\overline{B} = \overline{AR}(\overline{A})$?

12. Using I from Problem 11, $\overline{A}$ and $\overline{B}$ from Example 14.2.4, do we get the result $\overline{AR}(\overline{B^c}) = \overline{A^c}$?

13. If $\overline{A}$ is any triangular fuzzy number and I is the Gödel (Example 14.2.5) or the Gaines-Rescher (Problem 11) implication operator, then show that $\overline{AR}(\overline{A}) = \overline{B}$ for any fuzzy subsets $\overline{A}$ and $\overline{B}$ of the real numbers.

14. In Example 14.2.2 find $\overline{B'}(y)$ for other values of $\overline{A}(x_0)$, like 0.25, 0.5, 0.75.

15. Another method of finding $\overline{B'}$ from a given $\overline{A'}$, not using an implication operator from classical logic, is to define the fuzzy relation as

$$\overline{R}(x, y) = \overline{A}(x)\overline{B}(y).$$

    a. Given crisp input $\overline{A'}(x) = 1$ if $x = x_0$ and equals zero otherwise, find $\overline{B'}(y)$ if $\overline{A}(x_0) = 0, 0.5, 1$.

    b. Using t-norm min in equation (14.5) and this $\overline{R}$, is approximate reasoning consistent?

    c. Using this $\overline{R}$ and equation (14.5), t-norm min, does $\overline{AR}(\overline{B}^c) = \overline{A}^c$ hold?

16. If X and Y are finite it is usually easier to find $\overline{B'}$ given $x = \overline{A'}$. Let $X = \{x_1, \cdots, x_4\}$ and $Y = \{y_1, y_2, y_3\}$. Then $\overline{R}(x_i, y_j)$ can be shown as a 4×3 matrix. Let

$$\overline{A} = \{\frac{0.7}{x_1}, \frac{1}{x_2}, \frac{0.6}{x_2}, \frac{0.2}{x_4}\},$$

and

$$\overline{B} = \{\frac{0.5}{y_1}, \frac{0.8}{y_2}, \frac{1}{y_3}\}.$$

Form $\overline{R}$ using the Łukasiewicz I and call it $\overline{R_L}$. Also calculate $\overline{R}$ using the min operator of Example 14.2.3 and call it $\overline{R_m}$ and find $\overline{R}$ from the Gödel I of Example 14.2.5 and call it $\overline{R_G}$. Use t-norm min in equation (14.5).

    a. Which $\overline{R}$ makes approximate reasoning consistent?

b. Which $\overline{R}$ satisfies $\overline{AR}(\overline{B}^c) = \overline{A}^c$?

17. Rework Example 14.2.2 using the Reichenbach implication $I(a, b) = 1 - a + ab$.

18. How would you develop the theory of approximate reasoning if in equation (14.1) $\overline{A}$ and $\overline{B}$ were both type 2 (or both level 2) fuzzy sets?

19. In approximate reasoning $\overline{R}$ is a fuzzy relation. So let $\overline{R}$ be a fuzzy equivalence relation on $X \times Y$. In what ways will the theory of approximate reasoning change, if we use this $\overline{R}$, from approximate reasoning derived from a fuzzy relation computed from a crisp implication operator from classical logic?

20. In approximate reasoning we would like both equations (14.13) and (14.16) to hold. Consider $\overline{A}$ and $\overline{B}$ given and fixed. Can you then always solve equations (14.13) and (14.16) for a fuzzy relation $\overline{R}$ so that both equations are true? Use t-norm min in the compositions.

14.3 Multiple Rules

A block of fuzzy if-then rules is

$$if\ x\ is\ \overline{A}_i,\ then\ y\ is\ \overline{B}_i, \tag{14.18}$$

$1 \leq i \leq N$, where the $\overline{A}_i$ are fuzzy subsets of X and the $\overline{B}_i$ are fuzzy subsets of Y. Usually we need more than one rule to draw a reasonable conclusion. We would replace the single rule in equation (14.1) by a block of fuzzy rules:

$$\begin{aligned} if\ x\ is\ huge &,\ then\ y\ is\ very\ slow, \\ if\ x\ is\ very\ large &,\ then\ y\ is\ slow, \\ if\ x\ is\ large &,\ then\ y\ is\ medium, \\ if\ x\ is\ medium &,\ then\ y\ is\ medium, \\ if\ x\ is\ small &,\ then\ y\ is\ fast, \\ if\ x\ is\ tiny &,\ then\ y\ is\ very\ fast. \end{aligned} \tag{14.19}$$

Now given data on size $x = \overline{A'}$, what will be the conclusion for y being speed? All the terms *huge*, $\cdots$, *very fast* in the above block of rules are defined by the fuzzy numbers in Figures 14.1 and 14.2. After we substitute the $\overline{A}_i$ in Figure 14.1 for *huge*, $\cdots$, *tiny* and the $\overline{B}_i$ in Figure 14.2 for *very slow*, $\cdots$, *very fast*, then equation (14.19) becomes equation (14.18), $N = 6$.

There are two basic methods to evaluate a block of fuzzy rule in approximate reasoning given some data $x = \overline{A'}$: (1) FITA, or first infer and then aggregate, and and (2) FATI, or first aggregate and then infer. We first describe FITA.

First choose an implication operator I from classical logic and we will use the same I in all the rules. For each rule construct the fuzzy relation $\overline{R}_i(x,y) = I(\overline{A}_i(x), \overline{B}_i(y))$, $1 \leq i \leq N$. Given data $x = \overline{A'}$ execute (fire) each rule using equation (14.5) to obtain conclusion $y = \overline{B'}_i$, $1 \leq i \leq N$. Then aggregate the $\overline{B'}_i$ into a final conclusion $\overline{B^*}$. The aggregation operator could be union, defined by some t-conorm. We will assume union is the aggregation operation so that

$$\overline{B^*} = \bigcup_{i=1}^{N} \overline{B'}_i. \tag{14.20}$$

We may think of obtaining from each rule $\overline{B'}_1$ or $\cdots$ or $\overline{B'}_N$ and we model the "or" with union.

FATI first aggregates all the rules into one (super or global) rule. This single rule is modeled by a fuzzy relation $\overline{R}$ which is the aggregate of all the $\overline{R}_i$, $1 \leq i \leq N$. If we also aggregate using union then

$$\overline{R} = \bigcup_{i=1}^{N} \overline{R}_i. \tag{14.21}$$

Given data $x = \overline{A'}$ we only execute (fire) one rule given by $\overline{R}$ and we compute the final conclusion $\overline{B^*}$ using equation (14.5).

Let us abbreviate the method of going from $\overline{A'}$ to $\overline{B^*}$ under FITA as $\overline{B^*} = FITA(\overline{A'})$, a fuzzy function. Also, let $\overline{B^*} = FATI(\overline{A'})$ be the fuzzy function under FATI. Both methods depend on the implication operator I, the t-norm used in equation (14.5) and the method of aggregation. Let us assume that FITA and FATI both use the same I in all rules, both use the same t-norm in equation (14.5) and both employ union for aggregation. We now wonder: (1) what is the relationship, if any, between FITA and FATI?; (2) is either method consistent?; and (3) does either procedure satisfy $\overline{\Delta}(\overline{B}_i^c) = \overline{A}_i^c$, $1 \leq i \leq N$, for $\overline{\Delta}$ being FATI or FITA?

Example 14.3.1

Given some data on size $x = \overline{A'}$ we will evaluate the block of rules given in equation (14.19). We will first evaluate using a new method that will turn out to be equivalent to FATI. We need to put numbers into Figures 14.1 and 14.2: (1) let $a_1 = 10$, $a_2 = 20, \cdots, a_7 = 70$ kilograms for Figure 14.1; and (2) $b_1 = 20, b_2 = 40, b_3 = 60, b_4 = 80, b_5 = 100, b_6 = 120$ kph. For input let $\overline{A'} = (40/45/50)$.

We first find h_i the height of the intersection of $\overline{A'}$ and $\overline{A}_i$, or

$$h_i = \max_x\{\min[\overline{A'}(x), \overline{A}_i(x)]\}, \tag{14.22}$$

for $1 \leq i \leq 6$. Define

$$\overline{B'}_i = \min(h_i, \overline{B}_i), \tag{14.23}$$

for $1 \leq i \leq 6$ and

$$\overline{B^*} = \bigcup_{i=1}^{6} \overline{B'}_i. \tag{14.24}$$

We cut off the top of each $\overline{B}_i$ at height h_i and then union (using the t-conorm max) for all the results.

In this example we find that $h_1 = \cdots = h_4 = 0$ but $h_5 = h_6 = 2/3$. $\overline{B^*}$ is shown in Figure 14.3. If we need a crisp answer for speed, then defuzzify $\overline{B^*}$. To defuzzify a fuzzy set is to assign a representable crisp number to the fuzzy set. We could use the center of the core, or the center of gravity, as the defuzzified value. Defuzzification was discussed in Section 4.6 of Chapter 4.

Next consider the FATI discussed above. For each rule use the $\overline{R}$ in Example 14.2.3. That is

$$\overline{R}_s(x, y) = \min(\overline{A}_i(x), \overline{B}_j(y)). \tag{14.25}$$

Aggregate all the $\overline{R}_s$, $1 \leq s \leq 6$, using union (the t-conorm max) to fuzzy relation $\overline{R}$. Given $\overline{A'} = (40/45/50)$, use equation (14.5) to obtain $\overline{B'}$. Then $\overline{B'} = \overline{B^*}$ in Figure 14.3.

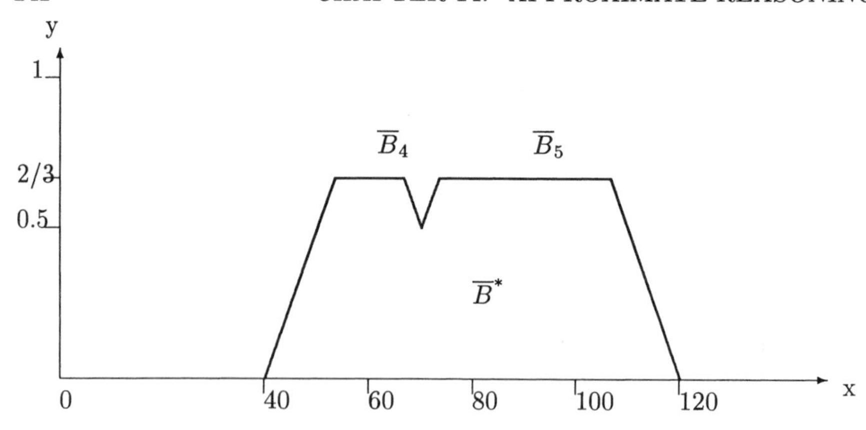

Figure 14.3: Conclusion in Example 14.3.1

14.3.1 Exercises

1. In Example 14.3.1 use the same $\overline{R}$, same data $\overline{A'}$, etc and find the final fuzzy conclusion using FITA. Is $FITA(\overline{A'}) \leq FATI(\overline{A'})$ or is $FATI(\overline{A'}) \leq FITA(\overline{A'})$?

2. Can FITA be consistent? Construct a block of two rules ($N = 2$) and find $\overline{A'}$ equal to $\overline{A}_1$ (or $\overline{A}_2$) so that $\overline{B^*} \neq \overline{B}_1$ (or $\overline{B}_2$). Use any implication operator used in Section 14.2. This will show that FITA may not be consistent.

3. Do the same as in Problem 2 for FATI.

4. Do the same as in Problem 2 to show that $FITA(\overline{B}_i^c)$ may not equal $\overline{A}_i^c$.

5. Do the same as in Problem 4 to show that $FATI(\overline{B}_i^c)$ may not equal $\overline{A}_i^c$.

6. In Example 14.3.1 show that $FATI(\overline{A'}) = \overline{B^*}$.

7. Design a HFNN (use the instructions to the problems in Section 13.3.1) for FITA.

8. Design a HFNN for FATI.

9. Let $X = \{x_1, \cdots, x_4\}$ and $Y = \{y_1, y_2, y_3\}$. We have three fuzzy rules "if $x = \overline{A}_i$, then $y = \overline{B}_i$", $1 \leq i \leq 3$, where the membership values of these fuzzy sets are:

   a. $\overline{A}_1$: 1, 0.7, 0.4, 0.2;
   b. $\overline{A}_2$: 0.2, 0.4, 1, 0.7;
   c. $\overline{A}_3$: 0.2, 0.4, 0.7, 1;

d. $\overline{B}_1$: 0.3, 0.8, 1;

e. $\overline{B}_2$: 0.3, 1, 0.8; and

f. $\overline{B}_3$: 1, 0.8, 0.3.

Use the $\overline{A'}$ given below to find $\overline{B^*}$ for I being the Łukasiewicz implication operator, for I being the Gödel implication and $\overline{R}$ equal min from Example 14.2.3. Also use the t-norm min in equation 14.5 and use the t-conorm max for union. Do both FATI and FITA and compare the results.

i. $\overline{A'} = (0.2, 0.4, 1, 0.7)$;

ii. $\overline{A'} = (0.5, 0.5, 0.5, 0.5)$; and

iii. $\overline{A'} = (1, 1, 1, 1)$.

10. We may summarize FITA as $\overline{B^*} = \bigcup(\overline{A'} \circ \overline{R}_i)$ and FATI as $\overline{B^*} = \overline{A'} \circ (\bigcup \overline{R}_i)$. Here we think of connecting the rules with "or" and using "union" to model "or". If we think of connecting the rules with "and", then we would aggregate using intersection. Let $\overline{B^*} = \bigcap(\overline{A'} \circ \overline{R}_i)$ and $\overline{B^*} = \overline{A'} \circ (\bigcap \overline{R}_i)$. Model union with C_m = max and intersection with $T_m = min$. Consider using the Łukasiewicz I, the Gödel I and the $\overline{R}$ equal min of Example 14.2.3. What relationships are there, if any, between these four methods of finding $\overline{B^*}$?

14.4 Discrete Case

Consider evaluating the block of rules in equation (14.18) when all the $\overline{A}_i$, $\overline{B}_i$ and data $\overline{A'}$ are continuous fuzzy numbers. In computer applications we would first discretize all fuzzy numbers. Suppose all the $\overline{A}_i$ and all the possible data values $\overline{A'}$ are continuous fuzzy numbers in an interval $[a_1, b_1]$. Choose a positive number M and numbers x_i so that $a_1 = x_1 < x_2 < \cdots < x_M = b_1$. Define $\overline{A}_i(x_j) = \alpha_{ij}$, $1 \leq i \leq N$, $1 \leq j \leq M$ and $\alpha'_j = \overline{A'}(x_j)$, $1 \leq j \leq M$ for the data. Let $\alpha_i = (\alpha_{i1}, \cdots, \alpha_{iM}), \alpha' = (\alpha'_1, \cdots, \alpha'_M)$.

Example 14.4.1

Let $\overline{A}_1 = (0.5/0.75/1.0)$ and $[a_1, b_1] = [0, 1]$. If M=11 and $0 = x_1 < 0.1 = x_2 < \cdots < x_{11} = 1.0$, then a discrete approximation to $\overline{A}_1$ is

$$\overline{A}_1 \approx \{\frac{\alpha_{11}}{x_1}, \frac{\alpha_{12}}{x_2}, \cdots, \frac{\alpha_{1,11}}{x_{11}}\}, \tag{14.26}$$

or

$$\overline{A}_1 \approx \{\frac{0}{x_1}, \cdots, \frac{0}{x_6}, \frac{0.4}{x_7}, \frac{0.8}{x_8}, \frac{0.8}{x_9}, \frac{0.4}{x_{10}}, \frac{0}{x_{11}}\}. \tag{14.27}$$

Notice that nowhere in the discrete approximation is the membership value equal to one. We can change the x_j values so that somewhere the membership is one, but for $N = 20$ and hundreds of possible values for $\overline{A'}$ we will be unable to choose the x_j so that all discrete approximations are normalized. The fact that some discrete approximations may not be normalized (no membership value equals one) will be important if we want to check consistency.

Now discretize the $\overline{B}_i, \overline{B'}_i$ and $\overline{B^*}$ of Section 14.3. Suppose all of these continuous fuzzy numbers are in the interval $[a_2, b_2]$. Choose positive integer P and y_j so that $a_2 = y_1 < y_2 < \cdots < y_P = b_2$. Let $\beta_{ij} = \overline{B}_i(y_j), \beta'_{ij} = \overline{B'}_i(y_j), \beta^*_j = \overline{B^*}(y_j)$ and $\beta_i = (\beta_{i1}, \cdots, \beta_{iP}), \beta'_i = (\beta'_{i1}, \cdots, \beta'_{iP})$ and $\beta^* = (\beta^*_1, \cdots, \beta^*_P)$. Using FITA, or FATI, the input is α' and the output is β^*.

First choose an implication operator I and compute an $M \times P$ (Type I) fuzzy matrix $\overline{R_a}$ whose elements r_{aij} in the interval $[0, 1]$ are

$$r_{aij} = I(\alpha_{ai}, \beta_{aj}), \tag{14.28}$$

for $1 \leq a \leq N$. We obtain a fuzzy relation $\overline{R}_a$ for each rule. Then equation (14.5) becomes

$$\alpha' \circ \overline{R}_a = \beta'_a, \tag{14.29}$$

$1 \leq a \leq N$. The composition "∘" in the above equation is the max-min composition. That is

$$\beta'_{aj} = \max_i \{\min\{\alpha'_i, r_{aij}\}\}, \tag{14.30}$$

for $1 \leq j \leq P$.

Now let us first look at FITA. Fire (execute) each rule given the data α' to obtain β'_a, $1 \leq a \leq N$, and aggregate using t-conorm max to get the final conclusion β^* whose components are

$$\beta^*_j = \max_{1 \leq a \leq N} \{\beta'_{aj}\}. \tag{14.31}$$

Under FATI we first aggregate all the $\overline{R}_a$ using max to get an $M \times P$ fuzzy matrix $\overline{R}$ whose elements are

$$r_{ij} = \max_{1 \leq a \leq N} \{r_{aij}\}. \tag{14.32}$$

The final conclusion is

$$\beta^* = \alpha' \circ \overline{R}. \tag{14.33}$$

So in FITA we have

$$\beta^* = \max_a \{\alpha' \circ \overline{R}_a\}, \tag{14.34}$$

but in FATI it is

$$\beta^* = \alpha' \circ \{\max_a \overline{R}_a\}. \tag{14.35}$$

From these two previous equations we can see that

$$FITA(\alpha') \leq FATI(\alpha'), \tag{14.36}$$

for all data α'.

Example 14.4.2

This example will use FATI, $N = 3$, $M = P = 11$ and $\overline{A}_1 = \overline{B}_3 = (0/0.2/0.4)$, $\overline{A}_2 = \overline{B}_2 = (0.3/0.5/0.7)$ and $\overline{A}_3 = \overline{B}_1 = (0.6/0.8/1)$. There are only three rules with linguistic interpretations $\overline{A}_1 = \overline{B}_3 =$ "low", $\overline{A}_2 = \overline{B}_2 =$ "medium" and $\overline{A}_3 = \overline{B}_1 =$ "high". So the first rule would be: "if x is *low*, then y is *high*".

All fuzzy sets belong to the same interval $[0, 1]$ with $x_1 = y_1 = 0$, $x_2 = y_2 = 0.1$, $\cdots$, $x_{11} = y_{11} = 1.0$. We will use $\overline{R}$ equal to min from Example 14.2.3. So we compute $\alpha_i, \alpha', \beta_i, \beta'_i$ and construct $\overline{R}_a$ for each rule. Each $\overline{R}_a$ is an 11×11 fuzzy matrix. For example $\overline{R}_1$ is mostly zero except for rows 2 through 4 and columns 8 through 10 where the values are either 0.5 or 1.0.

Also, $\alpha_1 = \beta_3 = (0, 0.5, 1, 0.5, \cdots, 0)$, $\alpha_2 = \beta_2 = (0,0,0,0,0.5,1,0.5,0,0,0,0)$ and $\alpha_3 = \beta_1 = (0, \cdots, 0, 0.5, 1, 0.5, 0)$.

Now form $\overline{R} = \max_a\{\overline{R}_a\}$. If we input $\overline{A'} = \overline{A}_1$, or $\alpha' = \alpha_1$, we easily see that $\beta^* = \beta_1$. Now input $\alpha' = (0,0,0,0.5,1,0.5,0,0,0,0,0)$ for $\overline{A'} = (0.2/0.4/0.6)$. Then

$$\beta^* = (0,0,0,0,0.5,0.5,0.5,0,0,0,0). \tag{14.37}$$

What, if any, linguistic term can you assign to this β^*? $\overline{A'}$ was chosen in between $\overline{A}_1$ and $\overline{A}_2$.

We may design a hybrid neural net for FITA and FATI. A hybrid neural net is similar to a hybrid fuzzy neural net discussed in Section 13.3 in Chapter 13 except for that the signals, weights and shift terms are all real numbers in a crisp hybrid neural net.

Figure 14.4 shows a hybrid neural net for FITA when $M = P = N = 2$. Recall, there is no training (learning) needed in a hybrid neural net because it is designed to compute the same output as FITA. All neurons have the identity transfer function.

The weights from the input neurons to the second layer are the elements in $\overline{R}_1$ and $\overline{R}_2$. We will use min in place of multiplication and max for addition. So the input to neuron #3 in the second layer is

$$\beta'_{21} = \max\{\min\{\alpha'_1, r_{211}\}, \min\{\alpha'_2, r_{221}\}\}, \tag{14.38}$$

which is also its output. So the second layer computes β'_{1j} and β'_{2j}, $j = 1, 2$. The weights to the output neurons are all equal to one and we multiply the signal by one but combine using max. The input (which is the same as its output) to neuron #1 in the output layer is

$$\beta'_1 = \max\{\beta'_{11}, \beta'_{21}\}. \tag{14.39}$$

A hybrid neural net for FITA would be used for fast parallel computation, especially when there are many rules in the block of rules and when there are multiple clauses in the antecedent part of the rules.

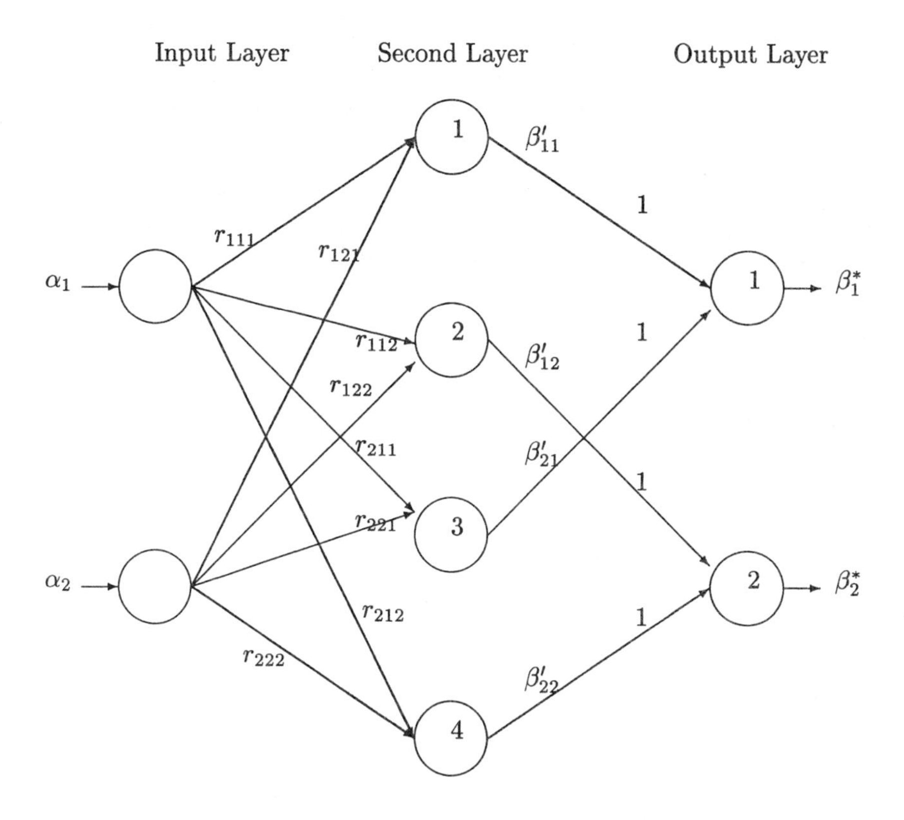

Figure 14.4: Hybrid Neural Net for FITA

14.4.1 Exercises

1. Determine the number of neurons needed in each layer in Figure 14.4 for arbitrary values of N, M and P.

2. Design a hybrid neural net for FATI when $M = N = P = 2$. Then determine the number of neurons needed in each layer for arbitrary values of M, N and P.

3. Suppose we have only one rule: "if $x = \overline{A}$, then y is $\overline{B}$ ". Assume we discretize $\overline{A}$ and all the possible data $\overline{A'}$ so that $\overline{A}(x_j) \neq 1$ for $1 \leq j \leq M$. Given data $x = \overline{A'} = \overline{A}$, show that the conclusion will not equal β, the discretized $\overline{B}$, so that approximate reasoning will not be consistent. Use any implication operator from Section 14.2.

4. Show that equation (14.36) is correct.

5. In Example 14.4.2 find β^* if the data is:

   a. $\overline{A'} = (0.5/0, 7/0.9)$, and

   b. $\overline{A'} = (0.1/0.5/0.9)$.

6. Rework Example 14.4.2 using the following implication operators:

   a. $I(a, b) = \min(1, 1 - a + b)$,

   b. $I(a, b) = 1$ if $a \leq b$ and it is zero otherwise, and

   c. $I(a, b) = 1$ if $a \leq b$ and it equals b otherwise.

7. Show how to simplify Example 14.4.2 if we have crisp input.

8. Rework Example 14.4.2 using FATI. Compare your answers.

9. How would FATI be simplified if we had crisp input?

10. We may also aggregate using min for intersection (Problem 10, Section 14.3.1). That is $\overline{B^*} = \cap\{\overline{A'} \circ \overline{R}_i\}$ or $\overline{B^*} = \overline{A'} \circ \{\cap \overline{R}_i\}$. Rework Example 14.4.2 this way for FATI and FITA.

11. Let $\overline{R}(x_i, y_j) = \overline{A}(x_i)\overline{B}(y_j)$ as in Problem 15 in Section 14.2.1. Use this method of finding the fuzzy relation for each rule in Example 14.4.2 and compare your results to those in Example 14.4.2.

12. This continues Problem 3. Assume that we discretize $\overline{B}$ to β so that $\beta_j \neq 1$ for all j. Now input $1 - \beta$ for $\overline{B}^c$. Use the implication operator I from Problem 11 in Section 14.2.1 (the Gaines-Rescher implication and t-norm T_m). Do we get for output $1 - \alpha$ (for $\overline{A}^c$)? This is the discrete version of the equation $\overline{AR}(\overline{B}^c) = \overline{A}^c$ studied in Section 14.2.

14.5 Other Methods

Approximate reasoning is not the only method of evaluating a fuzzy rule or a block of fuzzy rules. We saw that approximate reasoning has certain problems: (1) it may not be consistent; (2) $\overline{AR}(\overline{B}^c) = \overline{A}^c$ may fail to hold; (3) the situations pointed out in problems #2, #3 and #4 of Section 14.2.1; and (4) we may obtain a fuzzy conclusion (Figure 14.3) which is difficult to interpret. So, one may search for other methods of evaluating fuzzy rules which do not have these drawbacks. However, there are situations when we can not use approximate reasoning.

The linguistic variables Size and Speed defined in Section 14.2 are called numeric linguistic variables because their terms (members) can be specified by continuous fuzzy numbers (Figures 14.1 and 14.2). But there are also non-numeric linguistic variables. For non-numeric linguistic variables we can not define their terms by fuzzy numbers.

Example 14.5.1

An example of a non-numeric linguistic variable is "Mental Health" with members *major-depression, bipolar-disorder* and *schizophrenia.* We can not assign continuous fuzzy numbers, in some interval $[a, b]$, to these terms.

Consider a block of rules to determine the mental disorder of a patient:

1. if $(\cdots)$, then diagnosis = *major depression,*
2. if $(\cdots)$, then diagnosis = *bipolar-disorder,*
3. if $(\cdots)$, then diagnosis = *schizophrenia.*

The antecedent [the $(\cdots)$ part] of each rule is not important at this point. What the rules are to accomplish is to determine the membership values in the discrete fuzzy set

$$Diagnosis = \{\frac{\mu_1}{M-D}, \frac{\mu_2}{B-D}, \frac{\mu_3}{S}\}, \tag{14.40}$$

where $M - D$ is *major-depression,* $B - D$ is *bipolar-disorder* and S denotes *schizophrenia.* We want the block of rules to determine the μ_i in equation (14.40). Notice that there are no fuzzy sets in the conclusion part (the "then " part) of these rules. Hence, approximate reasoning can not be used.

The left side of the rules (the antecedent) may or may not contain fuzzy sets. Given some data , how do we evaluate this type of rule? Each clause in the antecedent (left side) is evaluated using the data and then we combine these values, using t-norms and t-conorms, into a final confidence value for the left side of the rule. This final confidence is then assigned to the conclusion part of the rule. For example, in the second rule in the block of rules given above, suppose this final antecedent confidence is 0.7. Then we assign 0.7 to the right side (bipolar-disorder) so that $\mu_2 = 0.7$ in the discrete fuzzy set called *Diagnosis* [equation (14.40)].

Example 14.5.2

Let the left hand side of the first rule, in the block of rules above, be: "if (patient has depressive symptoms) and (symptoms have lasted for at least one month), then $\cdots$". During an interview with the patient a doctor evaluates his/her degree of confidence in "has depressive symptoms" as 0.7 and evaluates "lasted for at least one month" as 1.0. The two clauses are connected by an "and" so we choose a t-norm T so that $T(0.7, 1)$ is the final confidence in the antecedent. If $T = T_m = \min$, then $T(0.7, 1) = 0.7$ and $\mu_1 = 0.7$ in equation (14.40).

So we see that this block of rules is evaluated completely different from approximate reasoning. The method discussed in this section is usually recommended for problems where we want to answer a question like "what is it". The possible answers are the terms in a discrete fuzzy set like for *Diagnosis* in equation (14.40).

Example 14.5.3

We want to decide what some object is that has been picked up on a radar screen. It could be a bird, a plane or a missile. Our classification depends on its Altitude, its Size and its Speed. Altitude, Size and Speed are all numeric linguistic variables whose terms are defined by continuous fuzzy numbers. The terms for Altitude will be: *very low, low, medium, high, very high* and *very-very high.* For Size we use Figure 14.1 and Figure 14.2 for Speed. There will be three primary blocks of rules: (1) the first block builds the discrete fuzzy for Altitude, (2) the second block produces the discrete fuzzy set Size, and (3) the last block of rules gives the discrete fuzzy set Speed. Using these three discrete fuzzy sets the final block of rules makes the discrete fuzzy set Classification which is

$$Classification = \{\frac{\mu_1}{bird}, \frac{\mu_2}{plane}, \frac{\mu_3}{missile}\}. \tag{14.41}$$

An example of a rule for Classification is: " if (size = *small*) and (altitude = *very high*) and (speed = *very fast*), then it is missile".

Given some data the four blocks of rules build the discrete fuzzy set Classification. The data is all crisp: real numbers for altitude, size and speed from the radar screen and other electronic devices.

14.5.1 Exercises

1. Finish the discussion in Example 14.5.3 which should include: (1) values for the a_i in Figure 14.1; (2) values for the b_i in Figure 14.2; (3) fuzzy numbers for the terms in Altitude as in Figures 14.1 and 14.2; (4) examples of the rules in each of the blocks of rules for Altitude, Size and Speed; (5) examples of rules for the classification of "bird" and "plane"; and (6) given crisp data x = altitude, y = size and z = speed, explain how all the rules are to be evaluated.

2. Design a rule based system to answer the following examples of "what is it?". Your discussion should follow that of Example 14.5.3 and Problem 1 above.

   a. Diagnosis of why your car will not start.

   b. Classification of a movie as G, PG, R or X.

3. Discuss other possible applications of this type of fuzzy reasoning.

Chapter 15

Genetic Algorithms

Consider an optimization problem where we want to find the maximum of a continuous function $y = f(x)$, $x = (x_1, \cdots, x_n)$, for $x \in D$, where D is a subset of $\mathbf{R^n}$. We may first try calculus methods where we solve the system of equations $\partial f / \partial x_i = 0$, $1 \leq i \leq n$, inside D, for all the critical points (the candidates for max or min). But this is an $n \times n$ system of non-linear (in general) equations to solve simultaneously for all solutions in D. This is, in general, a difficult problem to solve. But we will also have to look on the boundary of D, assuming that D is a closed subset of $\mathbf{R^n}$, for the max. This procedure only applies to functions f which are continuously differentiable in D. What would we do if f were not differentiable over all of D? What would we do to find the maximum of fuzzy functions $\overline{Y} = \overline{F}(\overline{X})$?

When calculus is too difficult to implement, or it is not applicable, we can try "search" methods. Random search is one such method. In this method we randomly generate M points x^i, $1 \leq i \leq M$, in D, trying to get some on the boundary of D when D is a closed set. Then we compute $y^i = f(x^i)$, $1 \leq i \leq M$, and keep the largest value of y^i, say y^l. That is, $y^l \geq y^i$, for all $i \neq l$. Our estimate of the maximum of f over D would then be y^l taken on at $x = x^l$. If M is sufficiently large, like in the millions, then we can get a good approximation to the true maximum of f on D. Random search is considered very inefficient compared to a directed search procedure. Genetic algorithms are a type of directed search.

At the first step of a genetic algorithm we randomly generate an initial population $\mathbf{P_0}$ whose members x^i are in D, $1 \leq i \leq M$. Now M can be relatively small, depending on the problem, like 500, 1000 or 2000. Next we evaluate $f(x^i) = y^i$, $1 \leq i \leq M$, to see how well we are doing in finding the maximum of f. Now select the m "best" x^i in $\mathbf{P_0}$ to generate the next population $\mathbf{P_1}$. In this problem by "best" we mean the m x^i values producing the m largest values of f. Set Q to be the set of m best x^i and rename them $q_1, \cdots, q_m$. Of course, $1 \leq m \leq M$, but m can be much smaller than M as $m = (0.1)M$ or $m = (0.5)M$.

We next describe how to get the next population $\mathbf{P_1}$ from $\mathbf{P_0}$ using crossover and mutation. Randomly choose two members of Q, say q_a and q_b, where

$$q_a = (q_{a1}, \cdots, q_{an}), \tag{15.1}$$

and

$$q_b = (q_{b1}, \cdots, q_{bn}). \tag{15.2}$$

Next randomly choose an integer k in $\{1, 2, \cdots, n\}$. For example, suppose $k = 7$. Then form q'_a and q'_b, two members of the temporary next population $\mathbf{P}'_1$, as follows

$$q'_a = (q_{a1}, \cdots, q_{a6}, q_{b7}, \cdots, q_{bn}), \tag{15.3}$$

and

$$q'_b = (q_{b1}, \cdots, q_{b6}, q_{a7}, \cdots, q_{an}). \tag{15.4}$$

This is the crossover operation. We interchange $(q_{a7}, \cdots, q_{an})$ and $(q_{b7}, \cdots, q_{bn})$. However, q'_a and q'_b may not belong to region D. If q'_a or q'_b do not belong to D, then discard them and do the crossover operation until both q'_a and q'_b are in D. Once q'_a and q'_b belong to D, place them in the temporary population $\mathbf{P}'_1$. Continue choosing two from Q until you have M members in $\mathbf{P}'_1$.

Next we have the operation of mutation. Randomly choose s members of $\mathbf{P}'_1$ and put them in set S and rename them $r_1, \cdots, r_s$. Normally s is small compared to M. The value of s could be $(0.01)M$, or $(0.001)M$. If this turns out to be less than one, then mutation may, or may not, be done. For example, if $M = 500$ and $s = (0.001)M = 0.5$, then we could do mutation every other population. For each r_i in S randomly generate an integer $k \in \{1, \cdots, n\}$. Suppose that $k = 4$. Randomly generate a real number x in some interval and replace r_{i4} with x giving

$$r'_i = (r_{i1}, r_{i2}, r_{i3}, x, r_{i5}, \cdots, r_{in}). \tag{15.5}$$

If r'_i is not in D, then repeat mutation until we get an r'_i in D. Then replace r_i in $\mathbf{P}'_1$ with r'_i. This produces the new population $\mathbf{P_1}$.

Check the stopping rule. If the stopping rule is fulfilled, stop the process and select the best [largest value of $f(x)$] x^i in $\mathbf{P_1}$ as our estimate of an $x \in D$ that maximizes f. If the stopping rule is not met, then replace $\mathbf{P_0}$ with $\mathbf{P_1}$ and continue to build the next generation.

There are a number of comments about the algorithm we need to make before going on to some examples. First the set D will reflect any constraints there are on the variables. In a constrained optimization problem there can be constraints like $x_i \geq 0$, for all i, or $x_1 < x_3$, or the sum of the x_i must equal one, etc. These constraints determine D. D is called the search space for the optimization problem.

The fitness function for a genetic algorithm determines how we choose the "best" m in a population to be used to generate the next generation (population). In the example discussed above the fitness function was obvious since it was a max problem. In some other problems it may be difficult to decide on a fitness function.

The stopping rule can be to terminate the iterations after K populations. Another stopping rule can be to end the iterations when the maximum value of f, over all members of a population, changes very little from generation to generation. However, when you stop the algorithm there is always the danger that it has converged to a local, not global, maximum (or minimum). To guard against this, run a number of trials. Each trial starts with randomly generating the initial population $\mathbf{P_0}$. Compare the results of the trials and if they all give essentially the same results, then you will have confidence in the results.

Also notice that we keep picking two members of Q to do crossover. Sometimes some members of Q will never be picked while others can be picked more than once. This means that the temporary population $\mathbf{P'_1}$ can be a mixture of "old" members of Q plus the new members q'_a and q'_b.

There are many types of crossover operations and we have described only one. Another type is the two point crossover. Randomly generate two integers k_1 and k_2 so that $1 \leq k_1 < k_2 \leq n$. Suppose $k_1 = 2$ and $k_2 = 6$. Then equations (15.3) and (15.4) become

$$q'_a = (q_{a1}, q_{b2}, \cdots, q_{b5}, q_{a6}, \cdots, q_{an}), \tag{15.6}$$

and

$$q'_b = (q_{b1}, q_{a2}, \cdots, q_{a5}, q_{b6}, \cdots, q_{bn}). \tag{15.7}$$

However, let us only use the one point crossover given by equations (15.3) and (15.4).

If $n = 1$, then we want to find the maximum of continuous $y = f(x)$ for x in some interval, say $x \in [a, b]$. Let p_i , $1 \leq i \leq M$, denote the population members of the initial population $\mathbf{P_0}$. Now $p_i = x$ for some x in $[a, b]$, for all i. But each p_i has only one element, or it is just a single real number, and there is no crossover. What we do when n is small, say $n \leq 5$, is to code each x_{ij} in $p_i = (x_{i1}, \cdots, x_{in})$ in binary notation (zeros and ones). For example if x=32, then $32 = (0)2^0 + (0)2^1 + (0)2^2 + (0)2^3 + (0)2^4 + (1)2^5$ so we write $x = (0, 0, 0, 0, 0, 1)$ in binary notation. As another example suppose x is in $[-100, 100]$ and we will use only two decimal places. Then $x = -56.78$ is first written as -5678 which is coded as $-(0, 1, 1, 1, 0, 1, 0, 0, 0, 1, 1, 0, 1)$ because $5676 = 2^1 + 2^2 + 2^3 + 2^5 + 2^9 + 2^{10} + 2^{12}$. Using binary notation each x_{ij} in p_i is translated into zeros and ones and concatenated to from p_i. Now crossover can be applied to each member of the initial population when n is small. However, we may not want to do this for all optimization problems.

Consider $n = 50$, $M = 5000$, and each $x_{ij} \in [-10,000, 10,000]$ and we use three decimal places. Calculate the length of each p_i in each population. They are over 1000 positions in length, and we need 5000 of them at each generation. We run into storage problems and the speed of the algorithm suffers.

Genetic algorithms are available from the internet. In your search engine try "genetic algorithm" or "evolutionary algorithm" to download the software. Evolutionary algorithm is a more advanced form of a genetic algorithm which emphasizes mutation over crossover.

Example 15.1

In this example we design a genetic algorithm to solve a fuzzy relational equation. This example will continue Example 7.5.2 of Chapter 7. We wish to find the r_2^* of equation (7.57) because the algorithm for r_2^{**} is not difficult to implement. Recall from Example 7.5.2 that it was easy to calculate r_i^{**} but much more difficult to obtain r_2^*. $\overline{S}$ was given in equation (7.60) and in that example we found that $r_2^{**} = (0, 0.8, 0.7, 0.5)$.

The initial population consists of $p = (r_1, r_2, r_3, r_4)$ so that $p \circ \overline{S} = (0.7, 0, 0.8, 0.5)$, which is the constraint on the problem. To cut down on the search space add the constraint $p \leq r_2^{**}$. The population size M should not be too large (say $M = 10$) because most randomly generated p will not satisfy the constraint (difficult to randomly generate a large initial population). The fitness function selects those p in a population that are minimal. A $v = (v_1, \cdots, v_4)$ in a population is minimal if there is no $u = (u_1, \cdots, u_4)$ in the population so that $u < v$, where $u < v$ means $u_i \leq v_i$, for all i, but at least one of the $\leq$ is a strict inequality ($<$). The size of Q and the value of m can vary from population to population. The stopping rule is when the population does not change from generation to generation. We may consider a number of trials for this problem.

Example 15.2

In this example we design a genetic algorithm to find the extension principle solution $\overline{X}_e$ to a system of fuzzy linear equations (see Chapter 11). First consider a $n \times n$ system of crisp linear equations written in matrix form

$$AX = B, \tag{15.8}$$

where $A = [a_{ij}]$ is an $n \times n$ matrix of real numbers, $X^t = (x_1, \cdots, x_n)$ is an $n \times 1$ vector of unknowns and $B^t = (b_1, \cdots, b_n)$ is an $n \times 1$ vector of real numbers. Let A_j be A with the jth column of A replaced by B. We write $\det(C)$ for the determinant of any $n \times n$ matrix C. We assume that

$\det(A) \neq 0$. Then the solution for X may be written as

$$x_i = \det(A_i)/\det(A), \tag{15.9}$$

for $1 \leq i \leq n$ from Cramer's rule. Let $\overline{X}_e^t = (\overline{X}_1, \cdots, \overline{X}_n)$ and $\overline{X}_i[\alpha] = [x_{i1}(\alpha), x_{i2}(\alpha)]$, $1 \leq i \leq n$. $\overline{X}_i$ is the fuzzification of equation (15.9). We see that [these are the same as equations (11.24) and (11.25) in Chapter 11]

$$x_{i1}(\alpha) = \min\{\frac{det(A_i)}{det(A)} | a_{ij} \in \overline{a}_{ij}[\alpha], b_i \in \overline{b}_i[\alpha]\}, \tag{15.10}$$

and

$$x_{i2}(\alpha) = \max\{\frac{det(A_i)}{det(A)} | a_{ij} \in \overline{a}_{ij}[\alpha], b_i \in \overline{b}_i[\alpha]\}, \tag{15.11}$$

for $1 \leq i \leq n$ and $0 \leq \alpha \leq 1$. We have also assumed that $\det(A) \neq 0$ for all $a_{ij} \in \overline{a}_{ij}[0]$. We now wish to design a genetic algorithm to estimate the values of equations (15.10) and (15.11) for selected values of α.

Choose a value for α. The members of any population are

$$p = (a_{11}, \cdots, a_{nn}, b_1, \cdots, b_n), \tag{15.12}$$

and the constraint is

$$a_{ij} \in \overline{a}_{ij}[\alpha], \ b_i \in \overline{b}_i[\alpha], \tag{15.13}$$

for all i and j.

Let us look at estimating $x_{i1}(\alpha)$. The fitness function is minimum [see equation (15.10)]. The constraint for crossover and mutation is equation (15.13). Of course, for each p in a population we have to compute $\det(A_i)/\det(A)$ to select the m best members for the set Q. For $x_{i2}(\alpha)$ the fitness function is maximum. Both problems can be run simultaneously.

Example 15.3

Here we want to design a genetic algorithm for training a (crisp) neural net. It will be the 2–3–1 neural net shown in Figure 13.1. We wish to train this neural net to approximate the values of some continuous function $z = f(x, y)$ for (x, y) in some region D in $\mathbf{R}^2$. First we need a training set. Choose (x_i, y_i) in D and compute $z_i = f(x_i, y_i)$, for $1 \leq i \leq K$. Without any explicit information on the behavior of f we would choose the data points (x_i, y_i) uniformly spread throughout D. The training set is $((x_i, y_i), z_i)$, $1 \leq i \leq K$. Given inputs x_i and y_i to the neural net let the output be $O_i = NN(x_i, y_i)$. The fitness function could be

$$\min(\sum_{i=1}^{K}(O_i - z_i)^2/K), \tag{15.14}$$

or

$$\min(\max_{1\leq i\leq K} |O_i - z_i|). \tag{15.15}$$

Population members are

$$p = (w_{11}, \cdots, w_{23}, u_1, u_2, u_3, \theta_1, \theta_2.\theta_3, \varphi). \tag{15.16}$$

The objective is to find values for the weights and shift terms so that the resulting neural net best fits the data (the training set), where best is measured by equation (15.14) or (15.15). Assume that we use the sigmoidal transfer function. The stopping rule would be when the objective function [equation (15.14) or (15.15)] gets sufficiently close to zero.

15.1 Exercises

1. In Example 15.1 we may get $m = 1$ for a certain population. What should we do then.

2. Because of the "tight" constraint in Example 15.1 we may have to alter the algorithm. Discuss changing each r_i in population member p to binary notation and using crossover only within each r_i value. Will this improve the search?

3. In Example 15.3 there was no constraint on the weights and shift terms. This means they can be any real number and the search space becomes all of $\mathbf{R}^{13}$. This search space is much to large. Discuss how to reduce the search space to some subset D of $\mathbf{R}^{13}$. Then this D becomes the constraint in the algorithm.

4. Consider a search space $D = [-B, B]^n$ whose elements are $x = (x_1, \cdots, x_n)$. Each x_i will be measured to d decimal places. Now code each x_i in binary notation and concatenate them into $p = (p_1, \cdots, p_L)$. Find the value of L if:

   a. $n = 100, B = 100, d = 3$; and

   b. $n = 20, B = 1000, d = 2$.

In the following problems you are asked to design a genetic algorithm for a certain problem. The initial population $\mathbf{P_0}$ of size M will contain members $p_i = (p_{i1}, \cdots, p_{in})$, $1 \leq i \leq M$. Describe the members of the initial population. Also describe any constraints there are on randomly generating the initial population. Define the fitness function. Also discuss, when needed, how you will compute values of the fitness function. Discuss any constraints there are on the crossover and mutation operations. Describe the stopping rule. Do you need binary notation? Any special considerations are there on the sizes on M, m and s? Do you need to change the crossover operation from the one described in the text?

5. To train a 2–N–1 neural net, with sign constraints, for Problem 1, Section 13.2.1, Chapter 13. Also explain how you get the training set.

6. Same as Problem 5 for Problem 2, Section 13.2.1.

7. Same as Problem 5 for Problem 3, Section 13.2.1.

8. Same as Problem 5 for Problem 4, Section 13.2.1.

9. Same as Problem 5 for Problem 5, Section 13.2.1.

10. To approximate the α-cuts of the extension principle solution $\overline{X}_e$ to:

    a. equations (5.34) and (5.35) of Chapter 5 (assume only real value solutions), and

    b. for the α-cuts of $\overline{\max}$ and $\overline{\min}$, Problem 6 in Section 4.4.1, if $\overline{M} = (1/2/3)$ and $\overline{N} = (2/3/4)$.

11. For the α-cuts of the fuzzy area of a fuzzy circle, equation (9.10), of Chapter 9.

12. For the α-cuts of the fuzzy perimeter of a fuzzy circle, Problem 12, Section 9.1.

13. For the α-cuts of the fuzzy area of a fuzzy triangle, Problem 13, Section 9.1.

Chapter 16

Fuzzy Optimization

16.1 Introduction

In fuzzy optimization we wish to maximize, or minimize, a fuzzy set (which is usually the value of a fuzzy function) subject to some fuzzy constraints. However, we can not maximize, or minimize, a fuzzy set, so we do what is commonly done in the area of finance where they wish to maximize (minimize) the value of a random variable whose values are restricted by a probability density function. To maximize $\overline{Z}$ we instead maximize the central value of $\overline{Z}$, maximize the area under the membership function to the right of the central value and minimize the area under the membership function to the left of the central value. This produces a multiobjective optimization problem subject to fuzzy constraints. We then change this multiobjective problem into a single crisp objective problem subject to the fuzzy constraints. To solve this final problem we propose to generate good approximate solutions using a genetic algorithm. This is all discussed in the next section. Other fuzzy optimization problems, including training a fuzzy neural net, solving fuzzy linear programming problems and fuzzy inventory control, using a genetic algorithm, are presented in the final section, section three.

16.2 Maximum/Mininimum of Fuzzy Functions

Consider the fuzzy function $\overline{Y} = \overline{F}(\overline{X}) = \overline{A} \cdot \overline{X}^2 + \overline{B} \cdot \overline{X} + \overline{C}$ for $\overline{A}, \overline{B}, \overline{C}$ being triangular fuzzy numbers and $\overline{X}$ a triangular shaped fuzzy number. We wish to find $\overline{X}$ to maximize (or minimize) $\overline{Y}$. It is clear, from calculus, what to do in the crisp case. If $y = f(x) = -2x^2 + 4x + 10$, then $dy/dx = -4x + 4 = 0$ at $x = 1$ and $\max f(x) = f(1) = 12$. However, we can not do this in the fuzzy case.

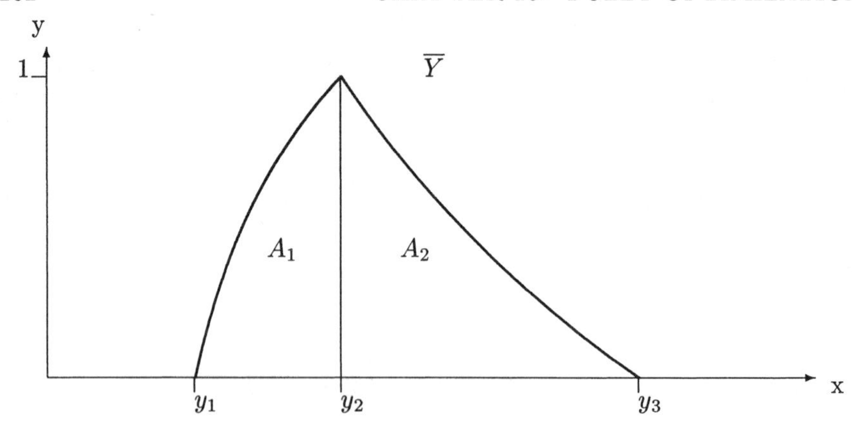

Figure 16.1: A Fuzzy Set to be Maximized

We can not maximize/minimize a collection of fuzzy sets like we can maximize/minimize a set of real numbers. The usual ordering of the real numbers gives a total (linear, complete) ordering and we can find the max/min of a finite set of real numbers and the supremum (infimum) of an infinite set which is bounded above (below). Most orderings of the fuzzy numbers are not a total ordering (see Section 4.5 of Chapter 4). The defuzzification method (Example 4.5.3) does give a total ordering for fuzzy numbers, but this ordering is not without problems (see Figure 4.15). Also, the defuzzification method only uses a central value of the fuzzy number and throws away all the other information in the fuzzy number.

The problem of finding the maximum/minimum of fuzzy $\overline{Y}$ is similar to a problem in finance where $\overline{Y}$ is the probability density function of a random variable Y that we wish to maximize (or minimize). If they wanted to max(Y), then they consider maximizing the expected value of Y, minimize the variance of Y and maximize the skewness to the right of the expected value. Now consider $\overline{Y}$ as in Figure 16.1. Let A_1 be the area under the graph from y_1 to y_2 and A_2 is the area from y_2 to y_3. So for max($\overline{Y}$) we will: (1) max(y_2), or maximize the central value of $\overline{Y}$; (2) max(A_2), or maximize the possibility of obtaining values more than y_2; and (3) minimize A_1, or minimize the possibility of values less than y_2. For min($\overline{Y}$) we would use: (1) min(y_2), (2) min(A_2), and (3) max(A_1).

The fuzzy optimization problem of max($\overline{Y}$) has become

$$[\min(A_1), \max(y_2), \max(A_2)], \tag{16.1}$$

where

$$\overline{Y} = \overline{A} \cdot \overline{X}^2 + \overline{B} \cdot \overline{X} + \overline{C}, \tag{16.2}$$

where $\overline{A}, \overline{B}, \overline{C}$ are given triangular fuzzy numbers and $\overline{X}$ is a triangular shaped fuzzy number.

Choose $M > 0$ sufficiently large so that $\min(A_1)$ is equivalent to $\max(M - A_1)$. So now we wish to $\max(M - A_1)$, $\max(y_2)$ and $\max(A_2)$. Obvious changes are for $\min(\overline{Y})$. Our final problem is

$$\max[\lambda_1(M - A_1) + \lambda_2 y_2 + \lambda_3 A_2], \tag{16.3}$$

subject to

$$\overline{Y} = \overline{A} \cdot \overline{X}^2 + \overline{B} \cdot \overline{X} + \overline{C}, \tag{16.4}$$

and $\lambda_i > 0$ for all i and $\lambda_1 + \lambda_2 + \lambda_3 = 1$. The λ_i are weights on the various objectives and are determined by the "decision maker". Suppose the decision maker [the one who wants to $\max(\overline{Y})$]) believes that $\max(y_2)$ and $\max(A_2)$ are equally most important. Then he/she might choose $\lambda_2 = \lambda_3 = 0.4$ and $\lambda_1 = 0.2$. Of course, different solutions can be generated for different choices of the λ_i. You might try a variety of values for the λ_i, always summing up to one, to see what you will obtain, show all these results to the decision maker and let he/she choose their best answer.

Example 16.2.1

Here we wish to find $\max(\overline{Y})$ if $\overline{Y} = \overline{F}(\overline{X}) = \overline{A} \cdot \overline{X}^2 + \overline{B} \cdot \overline{X} + \overline{C}$ for $\overline{A} = (-3/-2/-1), \overline{B} = (2/4/6), \overline{C} = (9/10/11)$ and $\overline{X} \approx (x_1/x_2/x_3)$. The crisp solution, using $\alpha = 1$ values, was found to be 12 at $x = 1$.

To reduce the search space for $\overline{X}$ we will assume that $\overline{X}$ belongs to the interval $[-10, 10]$. Let us use $\lambda_1 = 0.2, \lambda_2 = \lambda_3 = 0.4$ and $M = 1000$ in equation (16.3). The objective is

$$\max[0.2(1000 - A_1) + 0.4y_2 + 0.4A_2], \tag{16.5}$$

for

$$A_1 = \int_{y_1}^{y_2} \overline{Y}(x)dx, \tag{16.6}$$

$$A_2 = \int_{y_2}^{y_3} \overline{Y}(x)dx, \tag{16.7}$$

and

$$\overline{Y} = \overline{A} \cdot \overline{X}^2 + \overline{B} \cdot \overline{X} + \overline{C}. \tag{16.8}$$

We will solve for α-cuts of $\overline{X}$. Let $\overline{X}[\alpha] = [x_1(\alpha), x_2(\alpha)]$ and we will find $x_i(\alpha)$ for $\alpha = 0, 0.1, \cdots, 0.9, 1$, $i = 1, 2$. So we design a genetic algorithm (Chapter 15) to approximate the α-cuts of $\overline{X}$. The members of each population are

$$p = (x_1(0), x_1(0.1), \cdots, x_1(1), x_2(0.9), \cdots, x_2(0)). \tag{16.9}$$

Since we are using only eleven α-cuts of $\overline{X}$, we evaluate equation (16.8) via these α-cuts and obtain eleven α-cuts of $\overline{Y}$. The two areas (A_1, A_2) would be approximated using these α-cuts.

You are asked to finish the design of this genetic algorithm in the Exercises.

Example 16.2.2

This continues Example 8.4.2 of Chapter 8. There we wanted to find a value of $x \in [0, 2200]$ to

$$\max[\lambda_1(M - area_1(x)) + \lambda_2(\pi_2(x)) + \lambda_3(area_2(x))], \tag{16.10}$$

where

$$area_1(x) = (\pi_2(x) - \pi_1(x))/2, area_2(x) = (\pi_3(x) - \pi_2(x))/2,$$

$$\overline{\Pi}(x) = (\pi_1(x)/\pi_2(x)/\pi_3(x))$$

and

$$\overline{\Pi}(x) = (\overline{A} - \overline{B}x)x - (\overline{D}x + \overline{E}), \tag{16.11}$$

for $\overline{A} = (110/120/130)$, $\overline{B} = (0.03/0.04/0.05)$, $\overline{D} = (9/10/11)$ and $\overline{E} = (900/1000/1100)$. The value of x was restricted to the interval $[0, 2200]$ so that $\overline{A} - \overline{B}x \geq 0$.

Now design a genetic algorithm to find x in [0,2200] to maximize the expression in equation (16.10) for $\lambda_i = 1/3$ and for all i since the decision maker considers all goals are equal. Also choose $M = 1000$.

We would randomly generate the initial population whose members would be $p = x$ in [0,2200]. Since each p has only one member and crossover is undefined and you need to change to binary notation. Without binary coding all you have is mutation which ends up like a pure random search.

You are asked to finish the design of this genetic algorithm in the Exercises.

Example 16.2.3

We wish to $\min(\overline{Z})$ for

$$\overline{Z} = \overline{X}^2 + \overline{X} \cdot \overline{Y} + \overline{Y}^2 + 3\overline{X} - 3\overline{Y} + 4, \tag{16.12}$$

for $\overline{X}$ and $\overline{Y}$ triangular fuzzy numbers in $[-10, 10]$. We change the optimization problem to

$$\min[\lambda_1(M - A_1) + \lambda_2 z_2 + \lambda_3 A_3], \tag{16.13}$$

where $\overline{Z} \approx (z_1/z_2/z_3)$ and $A_1(A_2)$ is the area under the graph of $y = \overline{Z}(x)$ from z_1 to z_2 (z_2 to z_3) and the λ_i are positive adding up to one. Let us use $M = 1000$ and $\lambda_1 = \lambda_3 = 0.2$ and $\lambda_2 = 0.6$ since the decision maker feels the goal of minimizing z_2 is the most important objective.

We will use a genetic algorithm to search for $\overline{X}$ and $\overline{Y}$ in $[-10, 10]$ to minimize the expression in equation (16.13). Since $\overline{X}$ and $\overline{Y}$ are triangular let $\overline{X} = (x_1/x_2/x_3)$, $10 \leq x_1 < x_2 < x_3 \leq 10$ and $\overline{Y} = (y_1/y_2/y_3)$, $-10 \leq y_1 < y_2 < y_3 \leq 10$. So population members can be written as $p = (x_1, x_2, x_3, y_1, y_2, y_3)$. Use α-cuts and interval arithmetic to calculate $\overline{Z}$ with $\alpha = 0, 0.1, \cdots, 0.9, 1$. You are asked to finish the design of this genetic algorithm in the Exercises.

16.2.1 Exercises

The instructions about the design of genetic algorithms are the same as those for the Exercises in Chapter 15.

1. Finish the design of the genetic algorithm in Example 16.2.1.

2. Same as Problem 1 but for Example 16.2.2.

3. Same as Problem 1 but for Example 16.2.3.

4. Design a genetic algorithm for $\max(\overline{Y} = \overline{X}(1 - \overline{X}))$ for $\overline{X}$ being a triangular shaped fuzzy number in $[0, 1]$.

5. Extend the definition of fuzzy $\overline{\max}$, see Section 4.4 of Chapter 4, to $\overline{Z} = \overline{\max}\{\overline{A}_1, \cdots, \overline{A}_n\}$, $n \geq 3$ for all the $\overline{A}_i$ trapeziodal fuzzy numbers. Then design a genetic algorithm for $\overline{Z}$.

6. This continues Problem 4. Now we want to find $\overline{X} \approx (x_1/x_2/x_3)$ in $[0, 1]$ so that the fuzzy point $(\overline{X}, \overline{Y})$, where $\overline{Y} = \overline{X}(1 - \overline{X})$, is "closest" to the fuzzy point $(\overline{A}, \overline{B})$ for $\overline{A} = (0.2/0.3/0.4)$, $\overline{B} = (0.8/1/1.2)$. By closest we mean to minimize

$$[(D(\overline{A}, \overline{X}))^2 + (D(\overline{B}, \overline{Y}))^2]^{1/2},$$

   where $D(\overline{A}, \overline{X})$ is given in equation (3.80) in Chapter 3. Design a genetic algorithm for this problem.

7. Design a genetic algorithm to find the α-cuts of the substitute classical solution $\overline{X}_s$, using the distance measure D from equation (3.80) in Chapter 3, for:

   a. Problem 10, section 5.2.4;

   b. Problem 10, section 5.3.1;

   c. Problem 19, section 11.1.

8. Repeat Problem 7 using the distance measure D from equation (3.81) in Chapter 3.

9. Let

$$f(x) = \begin{cases} 1, & 0 \leq x < 1,\ 4 \leq x \leq 5 \\ 0, & 1 \leq x < 2,\ 3 \leq x < 4 \\ -1, & 2 \leq x < 3 \end{cases}$$

   and $\overline{A} = (1.5/2, 3/4.5)$. Using the extension principle extend f to $\overline{F}$ and let $\overline{Y} = \overline{F}(\overline{A})$. Find the α-cuts of $\overline{Y}$ for $\alpha = 0, 0.5, 1$. Since f is not continuous we can not use maximum/minimum of $\{f(x)|x \in \overline{A}[\alpha]\}$ for these α-cuts. Design a genetic algorithm for the $(\alpha = 0.5)$-cut.

10. Consider a fuzzy relational equation $\overline{R} \circ \overline{S} = \overline{T}$ that has no solution for $\overline{R}$. See Section 7.5 of Chapter 7. Consider a substitute solution $\overline{R}_s$, a fuzzy matrix, that solves the following minimization problem:

$$\min D(\overline{R} \circ \overline{S}, \overline{T}),$$

for some distance measure D between fuzzy matrices. First define a suitable D and then design a genetic algorithm to calculate $\overline{R}_s$ given $\overline{S}$ and $\overline{T}$.

16.3 Fuzzy Problems

There are fuzzy optimization problems that do not fit exactly into the models discussed in the previous section. We shall look at three such cases in this section.

Example 16.3.1

In this example we are going to train a fuzzy neural net. We want to design a genetic algorithm to train a Type III fuzzy neural net. A Type III fuzzy neural net was discussed in Section 13.3 of Chapter 13 and it has its inputs, weights and shift terms all fuzzy. Consider the 2–3–1 neural net shown in Figure 13.1 with inputs $\overline{X}_1$, $\overline{X}_2$, weights $\overline{W}_{ij}$, $\overline{U}_j$ and shift terms $\overline{\theta}_j$, $\overline{\varphi}$ all being continuous fuzzy numbers.

We want to train this fuzzy neural net to approximate the values of the fuzzy function $\overline{Z} = \overline{F}(\overline{X}, \overline{Y})$, (which is monotone increasing) for $\overline{X}$ in interval $[a, b]$ and $\overline{Y}$ in $[c, d]$. To obtain training data choose $\overline{X}_i$ in $[a, b]$ and $\overline{Y}_i$ in $[c, d]$ and compute $\overline{Z}_i = \overline{F}(\overline{X}_i, \overline{Y}_i)$, $1 \leq i \leq K$. We would usually choose the $\overline{X}_i$ ($\overline{Y}_i$) uniformly spread throughout $[a, b]$ ($[c, d]$).

We will input α-cuts of $\overline{X}_i$ and $\overline{Y}_i$ into the fuzzy neural net, do interval arithmetic within the net, and then the output will be intervals. Let the output from the fuzzy neural net be called $\overline{O}_i$ (different from $\overline{Y}$ in Section 13.3). We write $\overline{O}_i = FNN(\overline{X}_i, \overline{Y}_i)$, $1 \leq i \leq K$. So if the inputs are $\overline{X}_i[\alpha]$ and $\overline{Y}_i[\alpha]$, the output will be $\overline{O}_i[\alpha] = [o_{i1}(\alpha), o_{i2}(\alpha)]$, $1 \leq i \leq K$. Assume we do this for $\alpha = 0, 0.1, \cdots, 0.9, 1$. Let $\alpha_j = (j-1)/10$, $1 \leq j \leq 11$.

Now we need a fitness function, or some expression to minimize. Define

$$E = E_1 + E_2 + E_3, \tag{16.14}$$

$$E_1 = (1/K) \sum_{k=1}^{K} \sum_{j=1}^{10} (z_{k1}(\alpha_j) - o_{k1}(\alpha_j))^2, \tag{16.15}$$

$$E_2 = (1/K) \sum_{k=1}^{K} (z_{k1}(\alpha_{11}) - o_{k1}(\alpha_{11}))^2, \tag{16.16}$$

and

$$E_3 = (1/K) \sum_{k=1}^{K} \sum_{j=1}^{10} (z_{k2}(\alpha_j) - o_{k2}(\alpha_j))^2, \tag{16.17}$$

where $\overline{Z}_k[\alpha] = [z_{k1}(\alpha), z_{k2}(\alpha)]$.

We wish to find the fuzzy weights and shift terms to minimize E. The transfer function within each neuron in the second and output layer is the sigmoidal. Another possible fitness function is

$$E = \max\{E_1, E_2, E_3\}, \tag{16.18}$$

$$E_1 = \max_{1 \leq k \leq K} [\max_{1 \leq j \leq 10} |z_{k1}(\alpha_j) - o_{k1}(\alpha_j)|], \tag{16.19}$$

$$E_2 = \max_{1 \leq k \leq K} |z_{k1}(\alpha_{11}) - o_{k1}(\alpha_{11})|, \tag{16.20}$$

and

$$E_3 = \max_{1 \leq k \leq K} [\max_{1 \leq j \leq 10} |z_{k2}(\alpha_j) - o_{k2}(\alpha_j)|]. \tag{16.21}$$

Now we are ready to define members of each generation (population). Assume all the fuzzy weights and fuzzy shift terms are triangular fuzzy numbers. Also, the $\overline{X}_i$ and $\overline{Y}_i$ are all triangular so that $\overline{Z}_i$ and $\overline{O}_i$ will be triangular shaped fuzzy numbers. Then we may write: (1) $\overline{W}_{ij} = (w_{ij1}/w_{ij2}/w_{ij3})$; (2) $\overline{U}_j = (u_{j1}/u_{j2}/u_{j3})$; (3) $\overline{\theta}_j = (\theta_{j1}/\theta_{j2}/\theta_{j3})$; and (4) $\overline{\varphi} = (\varphi_1/\varphi_2/\varphi_3)$. Therefore, a population member p may be written as

$$p = (w_{111}, w_{112}, w_{113}, \cdots, \theta_{31}, \theta_{32}, \theta_{33}, \varphi_1, \varphi_2, \varphi_3). \tag{16.22}$$

To randomly generate the first population we need intervals for all the fuzzy weights and shift terms. Assume this has been done and all the $\overline{W}_{ij}$ belong to interval I_1, all the $\overline{U}_j$ are in I_2, the $\overline{\theta}_j$ belong to I_3 and $\overline{\varphi}$ is a fuzzy subset of I_4. Notice that each population member p describes a complete 2–3–1 fuzzy neural net. You are asked to complete the discussion of this genetic algorithm in the exercises.

Example 16.3.2

In this example we solve a fuzzy linear programming problem. We start with a crisp linear programming problem

$$\max(4x_1 + 5x_2) \tag{16.23}$$

subject to:

$$6x_1 + 3x_2 \leq 30, \tag{16.24}$$

$$3x_1 + 6x_2 \leq 30, \tag{16.25}$$

$$x_1, x_2 \geq 0. \tag{16.26}$$

All the values of (x_1, x_2) that satisfy the constraints (the inequalities) is called the feasible set. The feasible set for this problem is shown in Figure 16.2. The solution is at $(10/3, 10/3)$ with the maximum value of 30.

In fuzzy linear programming some of the parameters and variables can be fuzzy. We will look at the following fuzzy linear program

$$\max(\overline{Z} = \overline{c}_1 x_1 + \overline{c}_2 x_2) \tag{16.27}$$

subject to:

$$\overline{a}_{11} x_1 + \overline{a}_{12} x_2 \leq \overline{b}_1, \tag{16.28}$$

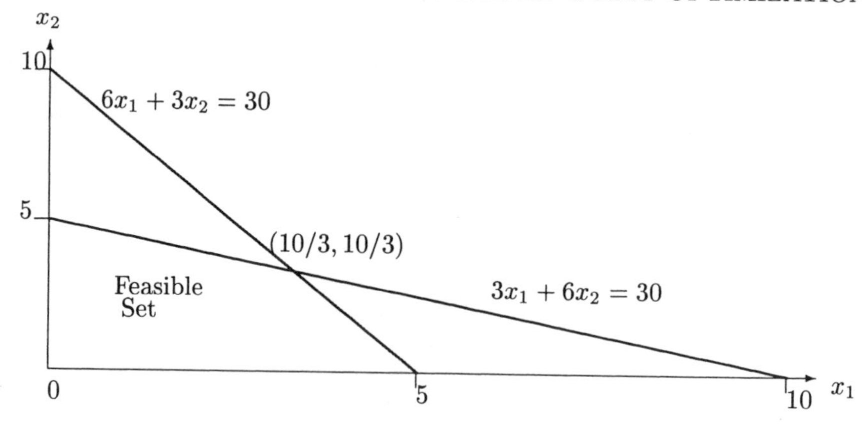

Figure 16.2: Feasible Set in Example 16.3.2

$$\overline{a}_{21}x_1 + \overline{a}_{22}x_2 \leq \overline{b}_2, \tag{16.29}$$

$$x_1, x_2 \geq 0, \tag{16.30}$$

where $\overline{c}_1, \overline{c}_2, \overline{a}_{11}, \cdots, \overline{a}_{22}, \overline{b}_1, \overline{b}_2$ are all triangular fuzzy numbers. The first thing we need to do is define the feasible set, or decide on what "$\leq$" we will use between fuzzy numbers. Let us use the "$\leq$" from Example 4.5.2 in Chapter 4. We write $\overline{M} \leq \overline{N}$ when $m_1(\alpha) \leq n_1(\alpha)$, $m_2(\alpha) \leq n_2(\alpha)$, $0 \leq \alpha \leq 1$, where $\overline{M}[\alpha] = [m_1(\alpha), m_2(\alpha)]$, $\overline{N}[\alpha] = [n_1(\alpha), n_2(\alpha)]$. Since the x_i are non-negative, the feasible set is all $x_i \geq 0$ so that

$$a_{111}(\alpha)x_1 + a_{121}(\alpha)x_2 \leq b_{11}(\alpha), \tag{16.31}$$

$$a_{112}(\alpha)x_1 + a_{122}(\alpha)x_2 \leq b_{12}(\alpha), \tag{16.32}$$

$$a_{211}(\alpha)x_1 + a_{221}(\alpha)x_2 \leq b_{21}(\alpha), \tag{16.33}$$

$$a_{212}(\alpha)x_1 + a_{222}(\alpha)x_2 \leq b_{22}(\alpha), \tag{16.34}$$

$0 \leq \alpha \leq 1$, where $\overline{a}_{ij}[\alpha] = [a_{ij1}(\alpha), a_{ij2}(\alpha)]$ and $\overline{b}_i[\alpha] = [b_{i1}(\alpha), b_{i2}(\alpha)]$.

Let us assume that $\overline{b}_i > 0$ so that at least $(x_1, x_2) = (0, 0)$ is feasible. That is, the feasible set is not empty.

Next we have to decide on how we are going to $\max(\overline{Z})$, $\overline{Z}$ is a triangular fuzzy number. Let $\overline{Z} = (z_1/z_2/z_3)$. We will do the same as in Section 16.2 (see Figure 16.1). The objective (fitness function) is to maximize

$$\lambda_1(M - A_1) + \lambda_2 z_2 + \lambda_3 A_2, \tag{16.35}$$

from equation (16.3), where $\lambda_i > 0$, their sum is one, A_1 is the area under the graph of the membership from z_1 to z_2 (which equals $(z_2 - z_1)/2$), etc.

Now design a genetic algorithm to find feasible (x_1, x_2) to maximize the expression in equation (16.35). To randomly generate the first population assume that the x_i belong to the interval $[0, 10]$. But the x_i must also be

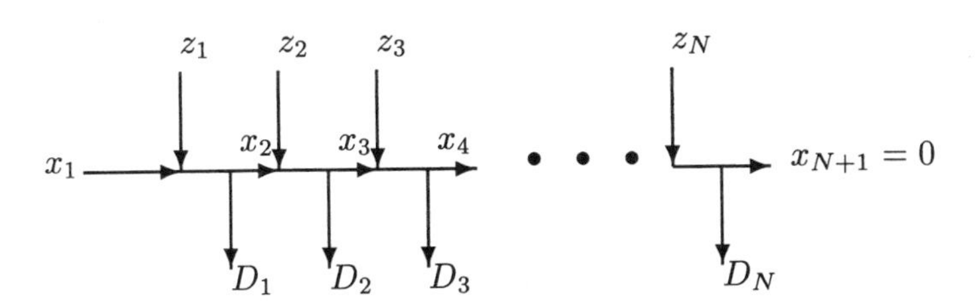

Figure 16.3: Inventory Problem

feasible [equations (16.31) - (16.34)]. Since a population member p has only two members (x_1, x_2) crossover may have a limited effect. Consider coding the x_i in binary notation. Choose values for the λ_i, let $M = 1000$. Also pick values for the $\overline{a}_{ij}, \overline{b}_i, \overline{c}_i$ all triangular fuzzy numbers whose vertex value is at the crisp number given in equations (16.23)–(16.25). You are asked to finish this discussion in the exercises.

Example 16.3.3

Solving a fuzzy inventory control problem. First let us describe the inventory problem we shall be studying. The flow is shown in Figure 16.3. The incoming inventory x_1 will always be a given real (not fuzzy) number. The variables are the $z_1, z_2, \cdots, z_N$, which are the amounts we are to order each period. The z_i, $1 \leq i \leq N$, will always be non-negative integers. If we allow the z_i to be fuzzy we will have to defuzzify them in the end, so we will start with them non-negative integers. The D_i represents the demand in the ith period and the x_i, $2 \leq i \leq N$ stand for the outgoing inventory, which will be the starting inventory for the next period. So we must have

$$x_{i+1} = x_i + z_i - D_i, \tag{16.36}$$

for $1 \leq i \leq N$.

There are only N periods and at the end of the planning horizon we assume that the final inventory will be zero. That is, we want $x_{N+1} = 0$. Other basic assumptions are: (1) no shortages are allowed, (2) zero delivery lag, and (3)instant replenishment at the start of each period. We wish to minimize the total inventory cost over the N periods. This cost will be made up of the following three components: (1) purchase cost, (2) ordering cost, and (3) holding cost.

We assume that we buy the item and we do not produce it ourselves. There may, or may not, be price breaks. A simple model, that we shall use is

$$c_i(z_i) = \begin{cases} 10z_i, & 0 \leq z_i \leq L_i, \\ 10L_i + 5(z_i - L_i), & z_i > L_i. \end{cases} \tag{16.37}$$

This means that in the ith period we pay \$10 per unit for the first L_i units and then pay \$5/unit for each additional unit. We will assume that these are all known numbers and they will not be fuzzy.

The ordering cost K_i in the ith period is the cost of placing the order, checking up on the order and putting the items into inventory when they arrive. This number is always difficult to estimate so we will model it using a fuzzy number. Then the total cost of obtaining z_i units at the start of the ith period is, for fuzzy $\overline{K}_i$,

$$\overline{C}_i = \begin{cases} 0, & z_i = 0, \\ \overline{K}_i + c_i(z_i), & z_i > 0, \end{cases} \tag{16.38}$$

The holding cost is assumed to be proportional to the ending inventory $x_{i+1} = x_i + z_i - D_i$. The model may be readily extended to cover any holding cost function $H_i(x_{i+1})$ by replacing x_{i+1} with $H_i(x_{i+1})$. For example, holding cost may be modeled as proportional to $(x_i + x_{i+1})/2$. Let h_i be the holding cost per unit for the ith period. This number, depending on interest on invested capital, depreciation, etc., is very difficult to determine exactly so it will be fuzzy in this problem. The holding cost for the ith period is

$$\overline{h}_i(x_i + z_i - D_i), \tag{16.39}$$

for fuzzy number $\overline{h}_i$ since the end inventory is $x_i + z_i - D_i$.

When no shortages are allowed let

$$\overline{TC}_i = \overline{C}_i(z_i) + \overline{h}_i(x_i + z_i - D_i), \tag{16.40}$$

and

$$\overline{Z} = \sum_{i=1}^{N} \overline{TC}_i. \tag{16.41}$$

We wish to find the z_i, $1 \leq i \leq N$, to minimize $\overline{Z}$. The constraint is $x_{N+1} = 0$.

Let us use positive trapezoidal fuzzy numbers for $\overline{K}_i$ and $\overline{h}_i$. So $\overline{Z} = (z_1/z_2, z_3/z_4)$ which will also be a trapezoidal fuzzy number. In place of $\min(\overline{Z})$ we use

$$\min(\lambda_1(M - A_1) + \lambda_2 \frac{z_2 + z_3}{2} + \lambda_3 A_2), \tag{16.42}$$

where $\lambda_i > 0$ and their sum is one, and $A_1(A_2)$ is the area under the graph of the membership function of $\overline{Z}$ from z_1 to $(z_2 + z_3)/2$ (from $(z_2 + z_3)/2$ to z_4).

To summarize, we wish to find non-negative integers z_i to minimize the expression in equation (16.42) subject to $x_{N+1} = 0$, given $x_1, c_i(z_i), N, D_i, \lambda_i, M$ all crisp and given $\overline{K}_i, \overline{h}_i$ fuzzy. Use α-cuts and interval arithmetic to evaluate all fuzzy equations.

Design a genetic algorithm to solve this problem. A population member p would be $(z_1, z_2, \cdots, z_N)$, but it must be feasible ($x_{N+1} = 0$). You are asked to finish the discussion of this algorithm in the exercises.

16.3.1 Exercises

The instructions for the design of a genetic algorithm are the same as those given in the exercises in Chapter 15.

1. Finish the discussion of the genetic algorithm for Example 16.3.1.

2. Explain what changes are needed in the genetic algorithm in Example 16.3.1 if:

   a. it is a Type I fuzzy neural net;

   b. it is a Type II fuzzy neural net;

   c. all the fuzzy weights, shift terms, $\overline{X}_i$, $\overline{Y}_i$ are trapezoidal fuzzy numbers and $\overline{Z}_i, \overline{O}_i$ are trapezoidal shaped fuzzy numbers; and

   d. all the fuzzy sets are just continuous fuzzy numbers.

3. Finish the discussion of the genetic algorithm in Example 16.3.2.

4. Explain what changes are needed in the genetic algorithm in Example 16.3.2 if:

   a. $\overline{X}_1$ and $\overline{X}_2$ are triangular shaped fuzzy numbers but all the parameters (the a_{ij}, c_i, b_i) are crisp; and

   b. $\overline{X}_1$ and $\overline{X}_2$ and all the parameters are triangular (shaped) fuzzy numbers.

5. Finish the design of the genetic algorithm in Example 16.3.3. Be sure you include how you check to see if a population member p is feasible.

6. What changes are needed in the genetic algorithm in Example 16.3.3 if:

   a. Demand is fuzzy?

   b. Demand is fuzzy and shortages are allowed?

7. Consider a fuzzy circle $\overline{C}$ given by $x^2 + y^2 = \overline{r}^2$, $\overline{r} = (0.5/1/1.5)$. Let $\overline{A}$ and $\overline{B}$ be two triangular fuzzy numbers and define $\overline{P} = \min(\overline{A}(x), \overline{B}(y))$ a fuzzy point. See Chapter 9 for fuzzy geometry. Assume $\overline{A}$ and $\overline{B}$ were chosen so that $\overline{P} \leq \overline{C}$. Now let $\overline{M} = (0.5/0.7/0.9)$ and $\overline{N} = (2/3/4)$, and set $\overline{Q} = \min(\overline{M}(x), \overline{N}(y))$. Design a genetic algorithm to find $\overline{P} \leq \overline{C}$ closest to $\overline{Q}$. By closest we mean to minimize

$$[(D(\overline{A}, \overline{M}))^2 + (D(\overline{B}, \overline{N}))^2]^{0.5},$$

   for D given by equation (3.80) of Chapter 3.

8. Suppose we have a block of rules

$$if\ x\ is\ \overline{A}_i,\ then\ y\ is\ \overline{B}_i,$$

$1 \leq i \leq N$. $\overline{A}_i$ are triangular fuzzy numbers all in interval $[a, b]$ and the $\overline{B}_i$ are also all triangular fuzzy numbers all in interval $[c, d]$. Let $\overline{B}^*$ be the conclusion using FITA (Section 14.3 of Chapter 14) using t-norm min, any implication operator I and t-conorm max for union. Defuzzify $\overline{B}^*$ to δ (see Section 4.6 of Chapter 4). Now suppose we have some data (x_i, z_i), $1 \leq i \leq m$, we want the fuzzy system to model. It could come from some continuous function. Also we assume the x_i are all in [a,b] and the z_i all belong to $[c, d]$. If the crisp input to the block of fuzzy rules is $x = x_i$, then let the defuzzified outcome be δ_i, $1 \leq i \leq m$. Design a genetic algorithm to find the $\overline{A}_i$ and $\overline{B}_i$, $1 \leq i \leq N$, to minimze

$$E = (1/m)\sum_{i=1}^{m}(z_i - \delta_i)^2.$$

9. Redo Problem 8 using FATI.

10. Let $\overline{P}_1(0,0)$ and $\overline{P}_3(2,4)$ be two fuzzy points which are right circular cones of base radius 0.2. See Chapter 9. Also $\overline{P}_2(2,0)$ and $\overline{P}_4(0,4)$ be two more fuzzy points which are right circular cones with base radius 0.4. Let $\overline{R}$ be the fuzzy rectangle defined by the $\overline{P}_i$, $1 \leq i \leq 4$. Now $\overline{R}^*$ is the closed fuzzy rectangle formed from $\overline{R}$ and $\overline{E}$ as follows

$$\overline{R}^* = \overline{R} \cup \overline{E},$$

where

$$\overline{E} = \begin{cases} 1, & 0 \leq x \leq 2,\ 0 \leq y \leq 4 \\ 0, & \text{otherwise} \end{cases}$$

Let $\overline{C}$ be the fuzzy circle $(x-a)^2 + (y-b)^2 = \overline{r}^2$, $\overline{r} = (r_1/r_2/r_3)$. Let the fuzzy area of $\overline{C}$ be $\overline{a}$. Design a genetic algorithm to find the fuzzy circle (find $a, b, \overline{r}$) of maximum area $\overline{a}$ inside $\overline{R}^*$ $(\overline{C} \leq \overline{R}^*)$.

Index

List of Figures

List of Tables

Druck:	Strauss Offsetdruck, Mörlenbach
Verarbeitung:	Schäffer, Grünstadt